MILESTONES IN THE DEVELOPMENT OF NATIONAL INFRASTRUCTURE FOR THE URANIUM PRODUCTION CYCLE

The following States are Members of the International Atomic Energy Agency:

AFGHANISTAN
ALBANIA
ALGERIA
ANGOLA
ANTIGUA AND BARBUDA
ARGENTINA
ARMENIA
AUSTRALIA
AUSTRIA
AZERBAIJAN
BAHAMAS
BAHRAIN
BANGLADESH
BARBADOS
BELARUS
BELGIUM
BELIZE
BENIN
BOLIVIA, PLURINATIONAL STATE OF
BOSNIA AND HERZEGOVINA
BOTSWANA
BRAZIL
BRUNEI DARUSSALAM
BULGARIA
BURKINA FASO
BURUNDI
CAMBODIA
CAMEROON
CANADA
CENTRAL AFRICAN REPUBLIC
CHAD
CHILE
CHINA
COLOMBIA
COMOROS
CONGO
COSTA RICA
CÔTE D'IVOIRE
CROATIA
CUBA
CYPRUS
CZECH REPUBLIC
DEMOCRATIC REPUBLIC OF THE CONGO
DENMARK
DJIBOUTI
DOMINICA
DOMINICAN REPUBLIC
ECUADOR
EGYPT
EL SALVADOR
ERITREA
ESTONIA
ESWATINI
ETHIOPIA
FIJI
FINLAND
FRANCE
GABON
GEORGIA
GERMANY
GHANA
GREECE
GRENADA
GUATEMALA
GUYANA
HAITI
HOLY SEE
HONDURAS
HUNGARY
ICELAND
INDIA
INDONESIA
IRAN, ISLAMIC REPUBLIC OF
IRAQ
IRELAND
ISRAEL
ITALY
JAMAICA
JAPAN
JORDAN
KAZAKHSTAN
KENYA
KOREA, REPUBLIC OF
KUWAIT
KYRGYZSTAN
LAO PEOPLE'S DEMOCRATIC REPUBLIC
LATVIA
LEBANON
LESOTHO
LIBERIA
LIBYA
LIECHTENSTEIN
LITHUANIA
LUXEMBOURG
MADAGASCAR
MALAWI
MALAYSIA
MALI
MALTA
MARSHALL ISLANDS
MAURITANIA
MAURITIUS
MEXICO
MONACO
MONGOLIA
MONTENEGRO
MOROCCO
MOZAMBIQUE
MYANMAR
NAMIBIA
NEPAL
NETHERLANDS
NEW ZEALAND
NICARAGUA
NIGER
NIGERIA
NORTH MACEDONIA
NORWAY
OMAN
PAKISTAN
PALAU
PANAMA
PAPUA NEW GUINEA
PARAGUAY
PERU
PHILIPPINES
POLAND
PORTUGAL
QATAR
REPUBLIC OF MOLDOVA
ROMANIA
RUSSIAN FEDERATION
RWANDA
SAINT KITTS AND NEVIS
SAINT LUCIA
SAINT VINCENT AND THE GRENADINES
SAMOA
SAN MARINO
SAUDI ARABIA
SENEGAL
SERBIA
SEYCHELLES
SIERRA LEONE
SINGAPORE
SLOVAKIA
SLOVENIA
SOUTH AFRICA
SPAIN
SRI LANKA
SUDAN
SWEDEN
SWITZERLAND
SYRIAN ARAB REPUBLIC
TAJIKISTAN
THAILAND
TOGO
TONGA
TRINIDAD AND TOBAGO
TUNISIA
TÜRKİYE
TURKMENISTAN
UGANDA
UKRAINE
UNITED ARAB EMIRATES
UNITED KINGDOM OF GREAT BRITAIN AND NORTHERN IRELAND
UNITED REPUBLIC OF TANZANIA
UNITED STATES OF AMERICA
URUGUAY
UZBEKISTAN
VANUATU
VENEZUELA, BOLIVARIAN REPUBLIC OF
VIET NAM
YEMEN
ZAMBIA
ZIMBABWE

The Agency's Statute was approved on 23 October 1956 by the Conference on the Statute of the IAEA held at United Nations Headquarters, New York; it entered into force on 29 July 1957. The Headquarters of the Agency are situated in Vienna. Its principal objective is "to accelerate and enlarge the contribution of atomic energy to peace, health and prosperity throughout the world".

IAEA NUCLEAR ENERGY SERIES No. NF-G-1.1

MILESTONES IN THE DEVELOPMENT OF NATIONAL INFRASTRUCTURE FOR THE URANIUM PRODUCTION CYCLE

INTERNATIONAL ATOMIC ENERGY AGENCY
VIENNA, 2023

Marketing and Sales Unit, Publishing Section
International Atomic Energy Agency
Vienna International Centre
PO Box 100
1400 Vienna, Austria
fax: +43 1 26007 22529
tel.: +43 1 2600 22417
email: sales.publications@iaea.org
www.iaea.org/publications

Printed by the IAEA in Austria
January 2023
STI/PUB/2019

IAEA Library Cataloguing in Publication Data

Names: International Atomic Energy Agency.
Title: Milestones in the development of national infrastructure for the uranium production cycle / International Atomic Energy Agency.
Description: Vienna : International Atomic Energy Agency, 2023. | Series: IAEA nuclear energy series, ISSN 1995–7807 ; no. NF-G-1.1 | Includes bibliographical references.
Identifiers: IAEAL 22-01555 | ISBN 978–92–0–128822–6 (paperback : alk. paper) | ISBN 978–92–0–128922–3 (pdf) | ISBN 978–92–0–129022–9 (epub)
Subjects: LCSH: Uranium. | Uranium — Production standards. | Uranium industry. | Uranium cycle (Biogeochemistry).
Classification: UDC 622.349.5 | STI/PUB/2019

FOREWORD

The IAEA's statutory role is to "seek to accelerate and enlarge the contribution of atomic energy to peace, health and prosperity throughout the world". Among other functions, the IAEA is authorized to "foster the exchange of scientific and technical information on peaceful uses of atomic energy". One way this is achieved is through a range of technical publications including the IAEA Nuclear Energy Series.

The IAEA Nuclear Energy Series comprises publications designed to further the use of nuclear technologies in support of sustainable development, to advance nuclear science and technology, catalyse innovation and build capacity to support the existing and expanded use of nuclear power and nuclear science applications. The publications include information covering all policy, technological and management aspects of the definition and implementation of activities involving the peaceful use of nuclear technology. While the guidance provided in IAEA Nuclear Energy Series publications does not constitute Member States' consensus, it has undergone internal peer review and been made available to Member States for comment prior to publication.

The IAEA safety standards establish fundamental principles, requirements and recommendations to ensure nuclear safety and serve as a global reference for protecting people and the environment from harmful effects of ionizing radiation.

When IAEA Nuclear Energy Series publications address safety, it is ensured that the IAEA safety standards are referred to as the current boundary conditions for the application of nuclear technology.

Energy is essential for development. Nearly every aspect of development — from reducing poverty and raising living standards to improving health care and industrial and agricultural productivity — requires access to energy sources. Current forecasts suggest that global electricity use will increase by 65–100% by 2030, with most of the growth in developing countries. Many IAEA Member States have expressed interest in introducing, or reintroducing, uranium mining and production activities to meet their own energy needs or those of other countries.

To introduce or reintroduce uranium mining and production, a wide range of factors need to be considered. This publication elaborates on the 'Milestones approach' to the uranium production cycle to assist Member States in adopting a systematic and measured approach to responsible uranium mining and processing. The guidance provided here is within the context of the IAEA's other publications on the development of the uranium production cycle, such as the IAEA Safety Standards Series.

The IAEA is grateful to the experts who contributed to this publication. The IAEA officers responsible for this publication were B. Moldovan and P. Woods of the Division of Nuclear Fuel Cycle and Waste Management.

EDITORIAL NOTE

CONTENTS

1. INTRODUCTION

1.1. BACKGROUND

A national uranium production programme is a complex undertaking that requires careful planning. A Member State that decides to support such a programme, through either national or foreign investment, needs to make a commitment that the uranium will be used for peaceful purposes. Furthermore, development of a national uranium production programme requires the establishment of sustainable national infrastructure that provides governmental, legislative, regulatory and industrial support for the lifetime of the programme. These aspects need to be based on accepted nuclear safety standards, security guidelines, safeguards requirements and international good practices. Decision makers, governmental organizations, regulatory bodies, academic institutions and industrial organizations need to be consulted to ensure that the required infrastructure is developed to sustain a national uranium production programme

This publication was developed to facilitate the assessment of progress in the development of infrastructure in a Member State considering a national uranium production programme. To enhance the IAEA's support to Member States in developing such a programme, this publication provides a detailed description of the 'Milestones approach' [1] for the stages of the nuclear fuel cycle [2] to help Member States to understand the various stages of knowledge and infrastructure required when they undertake exploration for uranium deposits. Support of uranium exploration by a Member State entails support of uranium mining and processing and requires the establishment of relevant legislation and regulations. If uranium deposits are found, the knowledge to evaluate and potentially develop them for mining and processing in a socially, financially and environmentally sound manner is required before committing to these activities.

All aspects of the uranium production cycle 'from cradle to grave' (e.g. from exploration to site remediation) need to be considered by Member States in a logical and systematic way when planning to mine and process uranium-bearing ore. Completion of activities associated with these aspects can be characterized as milestones along the road to sustainable development of a national uranium production programme. At the outset, the establishment of such a programme requires a systematic approach that can be divided into two general areas:

(a) Uranium exploration and resource evaluation. Applicable to all Member States.
(b) Uranium mining feasibility studies, engineering, construction, commissioning, mining, processing and closure. Applicable to Member

States that find one or more potentially significant uranium deposits, or where uranium is a potential by-product or co-product of the mining of other commodities, such as copper, gold, tin, rare earth elements, heavy mineral sands or phosphate.

Four milestones are identified for the uranium production cycle, each representing the beginning or boundary point of a stage or phase that a Member State may be currently advancing towards in the development of the uranium production cycle: exploration, development of the mine and processing facility, operation of the mine and processing facility, and finally decommissioning and remediation of the site. Sixteen aspects are identified at each phase and they need to be considered prior to advancing to the next milestone.

This publication can be used by Member States to assess their own status of uranium production development against each of the milestones. This includes the exploration, resource delineation, licensing, construction, commissioning and safe operation of a uranium mine and processing facility and, finally, the decommissioning and remediation phase. In addition, this publication aims to support Member States in regulating and overseeing uranium mining and processing activities. It may also be used to support self-assessment by a Member State already operating or wishing to restart a uranium mine and processing facility. This publication sets the foundation for IAEA integrated uranium production cycle review missions, which — upon request from the Member State — will review a Member State's progress in developing their national uranium production programme. Other stakeholders or interested parties, such as owners/operators (proponents), academic institutions, suppliers and contractors for uranium mining and processing, may also find this publication useful as they advance their respective programmes.

The information presented here is intended to relate the experience, lessons learned and good practices of countries with established uranium mines and processing facilities. Experience has shown that early attention to all the aspects presented in this publication can facilitate the efficient, safe and sustainable development and operation of a uranium mine and processing facility.

1.2. OBJECTIVE

This publication defines milestones in the development of the uranium production cycle and provides information on the activities that need to be carried out in a systematic manner at each milestone. A Member State can use it to ensure that it has achieved the following:

(a) Recognized the commitments and obligations associated with the establishment or re-establishment of a national uranium production programme;
(b) Prepared the local and national infrastructure adequately for the establishment or re-establishment of a national uranium production programme;
(c) Developed all the competences and capabilities required to regulate and potentially operate a national uranium production programme safely, securely and sustainably, and to manage the resulting waste.

1.3. SCOPE

The scope of this publication covers the governmental, regulatory and operational requirements to effectively and safely develop, commission and operate a uranium mine or processing facility. These requirements are considered from the time that a Member State decides to explore for uranium through to decommissioning and remediation, thereby encompassing the life cycle (cradle to grave) requirements.

The operation, waste management, and decommissioning and remediation of a uranium mine and processing facility are addressed to the degree necessary for planning purposes prior to advancing to operation. Good practice indicates that all key issues across the life cycle of a uranium project — including licensing, environmental assessment, construction, commissioning, operation, decommissioning, remediation and waste management — need to be considered early in the development of a uranium mine and processing facility. The related operational planning has to be well advanced prior to initiating any construction activities for the mine or processing facility. When the Member State is ready to commission a uranium mine or processing facility, it needs to have an understanding of the commitments required for the safe operation of these facilities and to have programmes in place that are sustainable for their life cycle through to their decommissioning, remediation and subsequent long term management, ensuring that they have 'started with the end in mind'.

This publication covers the milestones of the front end of the nuclear fuel cycle up to the point of production and transportation of uranium ore concentrate (UOC; e.g. yellow cake) and management of its waste. Refining, conversion and enrichment of uranium and nuclear fuel fabrication are outside the scope of this publication. It is not intended to be a comprehensive guide on feasibility studies and project management, but rather presents the national infrastructure requirements that need to exist at significant phases in the development process.

The main users of this publication are expected to be government decision makers and decision influencers, such as advisors in relevant government departments, regulatory bodies involved in the regulation of uranium mines and processing facilities, the uranium exploration and mining/processing industry, and researchers, including those in academic institutions.

1.4. STRUCTURE

This publication consists of three main sections, including the introduction. In Section 2, the four major milestones are presented, along with a brief description of each milestone. In Section 3, sixteen aspects of these four milestones are presented, along with the conditions required to achieve each milestone. The appendices provide two case studies.

2. MILESTONES IN THE DEVELOPMENT OF A URANIUM PROJECT

2.1. KEY CONCEPTS

A milestone describes a set of conditions that are expected to be met before advancing to a new phase in the development of a uranium project. The preparation of a Member State to introduce uranium exploration and potentially uranium mining and processing involves the completion of several activities, which can be divided into the following five progressive phases of development:

- — Phase 1: Development of a uranium exploration programme.
- — Phase 2: Exploration undertaken for the first time, or for the first time in many years, but with no significant commitment to proceed to mining and processing.
- — Phase 3: Initiation or reinvigoration of a uranium mining development with known exploitable uranium reserves.
- — Phase 4: Commissioning and operation, or increase of current capacity, of a uranium mine and processing facility.

— Phase 5: Uranium mines and processing facilities at the end of life, or mine sites being made safe but kept in a state suitable for possible reopening in the future.

2.2. THE MILESTONES

The completion of the infrastructure requirements prior to advancing to the next phase of development is marked by a specific milestone, at which progress and success of the development effort can be assessed and a decision made to advance to the next development phase. The four milestones in the uranium production cycle are the following:

— Milestone 1: Ready to make a commitment to explore for uranium.
— Milestone 2: Ready to commit to developing a uranium mine and processing facility.
— Milestone 3: Ready to operate a uranium mine and processing facility.
— Milestone 4: Ready to decommission and remediate a uranium mine and processing facility.

Following Milestone 4, and once all legal requirements have been met and verified during the post-decommissioning and remediation monitoring period, the project owner/operator has the right to apply to the regulatory body to be discharged of all further legal, financial and regulatory obligations of the project. If approved, the site would then be eligible to be part of an institutional control framework (this framework is outside the scope of this publication).

A schematic representation of the five phases and four milestones in the development, operation and decommissioning of a uranium mine and processing facility is provided in Fig. 1.

Like any other mineral derived raw material, uranium may or may not be technically, socially or economically viable to extract. Overall, mining is considered a temporary use of the land, with some operations running for 10–50 years or even longer. Following a successful decommissioning and remediation phase and agreement that the remediation has achieved the end state as approved by the regulatory body, it is expected that the lands will be returned for public or private use under a long term institutional control programme.

In the development of a national uranium mining and processing programme in a Member State, there are typically three major organizational entities involved. These are the government, the owner/operator (proponent or responsible party) of the uranium mine and processing facility, and the regulatory body. Each has a specific and independent role to play, with responsibilities changing as the

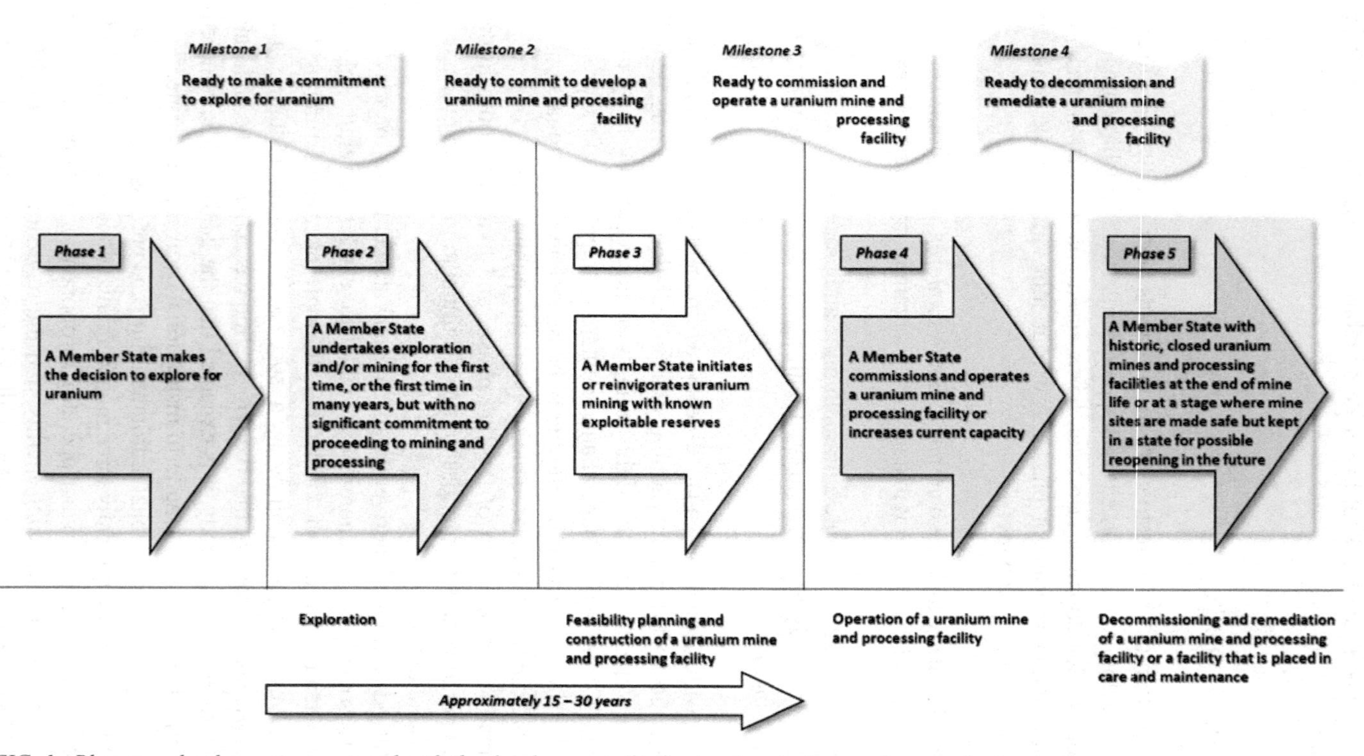

FIG. 1. Phases and milestones associated with the development of infrastructure in the uranium production cycle.

programme advances. It is assumed that the government is the entity that initiates exploration for uranium and development of a national infrastructure for uranium mining and processing through a well established national policy and strategy, as well as funding for these activities. This includes the development and funding of an independent regulatory body. The owner/operator may be state owned, another commercial entity or a combination of the two. The regulatory body needs to be effectively independent from the owner/operator and other government agencies responsible for the development of the uranium production programme but may exist within the government. Each of these entities is also accountable to the public, stakeholders and other interested parties and needs to be informed and consulted throughout the uranium production cycle.

For each milestone, 16 different aspects need to be considered. These are summarized in Table 1. The order of these aspects is not based on hierarchy or importance, as each aspect is important and requires careful consideration. The three main entities noted above (government, owner/operator and regulatory body) need to be aware of all these aspects and to manage them according to their respective roles and responsibilities.

2.2.1. Milestone 1: Ready to make a commitment to explore for uranium

This section describes the background and key considerations when planning for a uranium exploration programme (Phase 1) and the conditions that need to be met prior to initiating such a programme (Phase 2).

Each Member State interested in uranium exploration needs to have, or acquire, knowledge of the uranium potential of its geology. For the assessment of undiscovered uranium resources, both spatial/qualitative and quantitative approaches can be considered. Early in exploration, qualitative methods (e.g. literature studies, geologic mapping, geologic mineral surveys) are focused mainly on exploration targeting and project development. Late in exploration, quantitative methods (e.g. drilling, geochemical assaying, processing tests) are used for the assessment of potential recoverable mineral resources. The application of qualitative methods allows the efficient localization of the exploratory targets with greater chances of locating uranium deposits of a certain type. Quantitative methods are applied according to the geological knowledge and the degree of similarity of the deposits that could be found in a domain to determine the potential uranium ore grade and order of magnitude of uranium ore tonnage at the level of undiscovered resources. Grade–tonnage models and deposit density models of uranium deposits are required at this stage to complete this type of uranium potential modelling.

Irrespective of immediate uranium requirements, estimation of undiscovered uranium resources is valuable for sustaining the national policy on planning

TABLE 1. ASPECTS AND MILESTONES IN THE DEVELOPMENT OF A URANIUM PROGRAMME

Aspects	Conditions to achieve the milestone			
	Milestone 1	Milestone 2	Milestone 3	Milestone 4
National position	✓	✓	✓	✓
Safeguards	✓	✓	✓	✓
Legal and regulatory framework	✓	✓	✓	✓
Stakeholder engagement	✓	✓	✓	✓
Safety and radiation protection	✓	✓	✓	✓
Environmental protection	✓	✓	✓	✓
Protection/enhancement of cultural, tourism, farming, pastoral and related interests	✓	✓	✓	✓
Overview of the roles and responsibilities of the government, regulator and operator	✓	✓	✓	✓
Funding and financing	✓	✓	✓	✓
Security	✓	✓	✓	✓
Transport/export route	✓	✓	✓	✓
Human resources development	✓	✓	✓	✓
Site and supporting facilities (infrastructure)	✓	✓	✓	✓
Contingency planning	✓	✓	✓	✓

TABLE 1. ASPECTS AND MILESTONES IN THE DEVELOPMENT OF A URANIUM PROGRAMME (cont.)

Aspects	Conditions to achieve the milestone			
	Milestone 1	Milestone 2	Milestone 3	Milestone 4
Waste (including tailings) management and minimization	✓	✓	✓	✓
Industrial involvement including procurement	✓	✓	✓	✓

exploration and potential mine projects, in accordance with the adopted nuclear supply strategy. This supply strategy may take into account both the domestic supply for the manufacture of fuel to be used in nuclear reactors within the country and the delivery of UOC abroad. However, the magnitude and economics of the eventual emerging production projects need to be taken into account in order to evaluate the possibility of competing in the international market.

Exploration projects can lead to the discovery of uranium deposits and to the subsequent evaluation of the identified resources. This process can take several years, beginning with consultation and engagement of all interested parties and with the ultimate goal of obtaining and maintaining informed consent of the exploration activity. Owing to social, political, economic and technological factors, only a small fraction of the uranium resources identified advance to production. Therefore, expectations need to be managed carefully at every stage, especially with stakeholders and communities in project areas, through widely disseminated informational materials, consultations and public education campaigns.

The exploration process to confirm a new uranium deposit takes on average 10–15 years, from the moment that the very first indications are discovered to the confirmation of a potentially recoverable resource. Furthermore, exploration may continue throughout the life of the mining project to identify potential additional recoverable resources near the initial deposit. With site infrastructures already in place (e.g. processing plant, mine shops, waste management areas, access roads), the economics of finding another viable ore body near the existing mine become very attractive. In open pit operations, deep exploration drilling can be performed while production from the open pit is ongoing, and additional uranium resources can be added to the existing resources if results are favourable. Different phases of exploration can be considered, as outlined in the following sections.

2.2.1.1. *Selection of favourable areas for exploration*

There are many factors to be considered by a Member State when planning an exploration programme. These include geological factors, logistics and accessibility, environmental and social impact, land use, and economic and political issues. At the outset, political and geological factors are the most important. However, the other factors, which grow in significance if the project progresses, also need to be considered from the outset to ensure that an exploration programme results in the achievement of its objective, namely the development and exploitation of a mineral deposit. Additional elements of economic interest may also be identified, which may increase the viability of the project. Geological factors include knowledge of the local geology, including previous geophysical surveys (e.g. radiometric surveys), geochemistry, geomorphology, drilling and analytical logging data, and past production activities.

Historical information regarding the geology in a Member State can be obtained from government records, mining companies, universities, private exploration companies and publications from the IAEA and OECD Nuclear Energy Agency (NEA) [2–7]. Information obtained from exploration for minerals other than uranium may be used. For example, coal and oil exploration companies typically have radiometric information recorded in their drilling results. In Australia, the Olympic Dam project is essentially a copper mine, but its resource streams include additional mineral reserves of uranium, gold and silver.

2.2.1.2. *Exploration licence or permit*

An exploration licence or permit (they are the same at this stage in the uranium production cycle) provides the holder with the exclusive right to explore for a specified mineral group within the exploration licence area during the term of the licence. Prior to physically exploring for minerals, interested parties, first need to obtain an exploration licence/permit. The definitions and rules may differ between countries, and licences/permits are issued according to the mining laws of the host country. An exploration licence/permit does not allow mining, nor does it guarantee that a mining lease will be granted. Only a very small percentage of land that is subject to exploration licences/permits is developed into a mine. However, rules for the repartition of the stakes in a future project (between the investor and the country) can be included in the exploration licence/permit.

2.2.1.3. *Regional prospecting*

The objective of regional prospecting is to define the geological context of a selected area and potential zones for additional work. Regional prospecting includes geological mapping, remote sensing studies, airborne surveys, geochemical analysis and reconnaissance drilling to better define local geology. Regional exploration needs to focus on geological areas that have the potential to host uranium deposits. Selection of geological areas for more detailed exploration needs to be based on positive results from detailed and comprehensive analysis of all available geological, geophysical, geochemical and remote sensing data. The main activities at this phase of exploration are the identification of potential uranium hosting areas, the staking of claims and the application for relevant exploration licences/permits.

2.2.1.4. Detailed exploration

Once areas favourable for uranium mineralization, including ore grade or near ore grade mineralization, have been identified, the next stage of exploration may begin. In general, detailed exploration includes activities such as geological surveys, radiometric mapping, geochemical analysis, geophysical studies and drilling. Clear guidance needs to be provided to the managers of the exploration programme to protect workers and the environment. In most cases, ground geophysical surveys and drilling are essential to advance exploration at this stage. Detailed exploration involves a stage gate decision process: the potential of the zone for uranium mineralization is evaluated and a decision is made on whether to proceed to resource delineation and estimation or to remediate and vacate the exploration area.

In addition, during the first stage of detailed exploration, an environmental baseline study needs to be considered if there is a possibility that the project may proceed. The assessment must evaluate the baseline conditions of the site to support the determination of the anticipated impacts on the flora, fauna, wildlife and economy and to assess relevant historical and social factors. This is particularly important should the project advance to the mining stage. The preliminary baseline information that needs to be collected includes site location, meteorology, surface hydrology, hydrogeology (e.g. water quality, aquifer properties), flora and fauna, wildlife, soil and subsoil, background radiological characteristics, background non-radiological characteristics (e.g. heavy metals, pollutants), previous and current industrial and agricultural activities, local population, employment opportunities and other environmental features. This assessment needs to be conducted in consultation with local organizations and communities [8–10].

2.2.1.5. *Delineation drilling*

The delineation drilling stage of an exploration programme begins when the potential for significant resources has been recognized (during the detailed exploration stage) and a decision has been made to fully evaluate the prospect and accurately determine the resources. At this stage, it is essential to drill holes on a well defined grid pattern so an accurate estimation of resources can be made. The spacing of the grid pattern depends on the nature of the mineralization and, in particular, its spatial continuity. The spacing of delineation drill holes is also dependent on the degree of confidence that is required before a decision to begin mining can be made. If the deposit is only marginally economical, then the resources need to be determined very accurately and the drill hole spacing may need to be very small. Activities at this stage may include detailed geophysical, geological and geochemical analysis, topographical analysis, detailed drilling and logging, chemical analysis of drill core samples or drill cuttings, resource estimation modelling, mining tests and hydrometallurgical process evaluation tests (at laboratory and pilot plant scale). Increased regulatory oversight and controls are also common during this phase, and once again clear guidance to the managers of the exploration programme needs to be provided to protect workers and the environment. The outcome of the delineation drilling stage is a well defined uranium deposit with mineral resources and/or ore reserves if results are favourable. Expansion of the environmental baseline studies, the environmental impact statement (EIS), may also be required, as localized impacts from the expanded delineation drilling can occur.

2.2.1.6. *Resource estimation*

Resource estimation is an ongoing activity throughout the life of a mine, starting at exploration and continuing through development and production. The decision on whether to develop a mine to extract uranium from the defined deposit is made at this stage. Mineral resource and reserve classification are assigned to mineral deposits on the basis of their geological certainty and economic value. Because classification is an economic function, it is governed by statutes, regulations and industry best practice norms. There are several classification schemes globally, which are aligned with the International Committee for Mineral Reserves International Reporting Standards (CRIRSCO) code [11] and the OECD/NEA–IAEA classification scheme for uranium resources [2], such as the following:

(a) The Canadian Institute of Mining classification, or National Instrument (NI) 43–101 [12];

(b) The Australasian code for reporting exploration results, mineral resources and ore reserves, or JORC (Joint Ore Reserves Committee) Code [13];
(c) The South African code for reporting mineral resources and mineral reserves, or SAMREC (South African Mineral Resource Committee) Code [14].

2.2.1.7. Reporting

Reporting requires accuracy, reliability and transparency in the information derived from exploration results, resources and reserves. Many developing countries do not utilize national codes for reporting mining project data and results, and further action is required to establish legislation and a regulatory framework for reporting, as well as capacity building in the areas of administration and infrastructure (e.g. qualification committee, professional registry of competent persons). In contrast, publicly traded uranium exploration and mining companies usually report project deliverables using a codified set of rules and guidelines for reporting and displaying information related to mineral properties. Examples from Australia (JORC Code [13]), Canada (NI 43–101 [12]) and South Africa (SAMREC Code [14]) align with the CRIRSCO Code [11]. A reporting scheme specifically for uranium resources has been developed by OECD/NEA and IAEA [2] and is used by many Member States.

2.2.2. Milestone 2: Ready to commit to developing a uranium mine and processing facility

In preparing the infrastructure to initiate or reinvigorate uranium mining and processing (e.g. Phase 3) following identification of a uranium resource, there are several sequential activities that need to be completed. These include the following:

(a) Understanding the ore body and surrounding host material;
(b) Understanding the environmental conditions;
(c) Development of the mine and processing facility plans;
(d) Development of an infrastructure and services plan;
(e) Preparing an application for a licence to construct and a licence to operate;
(f) Construction of the mine and processing facility;
(g) Commissioning;
(h) Understanding the decommissioning and remediation requirements.

2.2.2.1. *Understanding the ore body and surrounding host material*

The first stage of mine development involves gaining an understanding of the ore body and its surrounding host material. This is accomplished through additional delineation drilling, which provides information on the depth, spatial geometric layout and hydrogeological conditions of the deposit. From these data a decision can be made as to whether to advance with underground mining, open pit (strip) mining or in situ recovery [15–17]. This drilling programme also provides information on ground stability and dewatering requirements. Finally, it provides information on the amount of mine rock that will be generated during the development and mining phases. Adequate segregation and management, including storage and treatment where appropriate, of mine rock material (including radiologically free clean rock and mineralized radioactive (contaminated) waste rock) from a safety and environmental perspective also need to be considered. Clean mine rock is a valuable construction material, and this asset needs to be identified early in the process.

From a processing perspective, the delineation drilling programme provides spatial information on the uranium grade of the ore deposit and its mineralogical and geochemical nature. The uranium grade and geochemistry of the ore are required to determine the processing and tailings management method that will be employed to extract and produce a marketable uranium concentrate.

This knowledge base may already be available for a pre-existing mine and processing facility developed during previous mining campaigns. From a due diligence perspective, however, such as is likely to be required by an investor or lender, a comprehensive review of this updated information is necessary to assess any changes in ground and hydrogeological conditions. In addition, a detailed review of any existing mine plan is required, and additional drilling may be necessary to confirm the reserves and resources and to verify that the current mine plan is still accurate and meets current safety and regulatory requirements.

2.2.2.2. *Understanding the environmental conditions*

The second stage of mine development involves acquiring a comprehensive understanding of the local environment and the potential impacts that mining activity could have on the local biota. This may be summarized in an EIS, an environmental and social impact study (ESIS), an environmental impact assessment (EIA), an environmental and social impact assessment (ESIA), or similar [8–10]. During this phase, the activities move towards a more comprehensive EIS as the extent and possible duration of the project are further defined.

The IAEA Safety Standards Series No. GSG-10, Prospective Radiological Environmental Impact Assessment for Facilities and Activities [18] states:

> "In the framework of international legal instruments or national laws and regulations, States may also require that, for some facilities and activities, a governmental decision making process, including a comprehensive initial assessment of possible significant effects on the environment, be carried out at an early stage in the development of the facility or activity. In this case, the radiological environmental impact assessment is generally part of a broader impact assessment, which is generally referred to as an 'environmental impact assessment' or by its common abbreviation EIA. An environmental impact assessment prospectively evaluates biophysical impacts (including radiological impacts) and also covers social, economic and other relevant impacts of a proposed activity or facility prior to major decisions being taken."

> "In the context of this Safety Guide, the term 'governmental decision making process' refers to the procedures carried out at all planning, pre-operational, operational and decommissioning stages by the government or governmental agencies, including the regulatory body, in deciding whether a project for a facility or an activity may be undertaken, continued, changed or stopped."

Therefore, an improved understanding of pre-existing conditions forms part of the environmental baseline study, which identifies the condition of, for example, the water courses, groundwater, transported dust, wildlife, biota, flora and fauna. Project stakeholders, including the authorities and those close to or dependent on the water, biota, wildlife, flora and fauna in the region, need to be consulted on the possible implementation of the uranium production project. Timely engagement of all stakeholders early in the project, starting with the exploration phase, is recommended. Obtaining social acceptance of the project may be the longest step in the study phase of any mining project.

Both water and waste management need to be included in an EIA. In keeping with all mining projects, it is necessary to consider water as a critical resource in terms of usage and overall management, including treatment and — where possible and as approved by the regulatory body — disposal. Consideration needs to be given to maximizing efficiencies for water use in mining and processing and that clean waters are not unnecessarily contaminated by mining or processing activities. Waste, such as that derived from stripping any overburden from the ore, needs to be characterized so that its location, either temporary or final, can be identified. In addition, decommissioning and mine closure activities and costs need to be considered at the initial stages of mine development as part of a full life cycle analysis.

It is important for the operator to develop end-of-life plans for the mine at this stage and before operations commence. Aspects to consider in the end of mine life plan include: decommissioning and remediation objectives and costs; desired end states; and future land use options, including long term institutional control, if appropriate.

2.2.2.3. Development of mine and processing facility plans

Once the resources and reserves have been delineated and an understanding of the structural geology and ore deposit has been gained, the next step is to develop a detailed mine plan. Also, environmental baseline conditions and preliminary environmental impacts need to be assessed at this point. The mine plan includes the type of mining proposed, the development and infrastructure requirements, and the dewatering and hydrometallurgical processes. Depending on the type of mining proposed, specific safety and training programmes need to be developed to ensure the safety of the workers and the general public. Some considerations include ground stability, ventilation, dust control, radiation safety (monitoring and management of gamma, alpha, long lived radioactive dust), electrical safety, conventional construction and operational safety, and safe operation of mining, transport and processing equipment. The mine plan also needs to include an understanding of the skilled workers required to manage and operate the mine. This can have an impact on the schedule of the project should extensive training be required prior to developing, commissioning and ultimately operating the mine.

Based on the type of mine (e.g. in situ recovery (ISR), underground, open pit) and processing facility proposed, a detailed engineering and construction plan needs to be developed. This includes the mine workings and associated infrastructure. For a hydrometallurgical processing facility, the ore mineralogy and geochemistry, as well as pilot plant test work, determine the processing options. The mine and processing facility construction plans need to be developed by a multidisciplinary team that includes geologists and mining, processing, civil, mechanical, environmental and electrical engineers, as well as a project management team to develop project, scope, budget, schedule, procurement, commissioning and startup plans [19–22].

2.2.2.4. Development of the infrastructure and services plan

Infrastructure and service requirements, including procurement, also need to be considered during the planning stage of the mine. These include, but are not limited to, the capability of the local electrical grid to support mining activities, access to water, roads, emergency response, administration offices, maintenance, warehousing and worker residences (camp facility), if required. The ability to readily procure equipment, spare parts, bulk reagents and fuels also need to be taken into consideration.

2.2.2.5. *Application for a licence to construct*

Applicable regulatory approvals need to be requested prior to advancing the mining project to the construction phase or restarting an existing mine. This may include formal public meetings to provide the public, non-government organizations, regulators and other interested stakeholders an opportunity to participate, provide feedback and ask questions on safety, environmental and socioeconomic aspects (EIA, EIS) prior to approval of a mine project. The entire process, from resource delineation through to regulatory hearings, may take five to ten years to complete owing to the complex nature of each of the phases of mine and processing facility development.

At this stage, the Member State needs to have a regulatory framework developed, including all necessary policies, standard operating procedures and related regulatory oversight and reporting frameworks for the construction and eventual operation of the facility. This includes aspects such as radiation protection, conventional safety and waste management. In addition, the Member State needs to have environmental regulations that require the operator to meet regulatory requirements for environmental performance that are in keeping with the best available and practical technology. There also needs to be guidelines and regulations in place for management systems such as human resources development (e.g. recruitment, training), information knowledge management and contractor management to ensure safe, reliable production. This infrastructure would be expected to be in compliance with international standards and would cover all current activities, practices and facilities in that Member State [8, 21].

2.2.2.6. *Construction of mine and processing facility*

Once regulatory approval has been granted to construct the uranium mine and processing facility, construction may begin. Construction is a structured, regimented process. The owner/operator may contract a specialized company or numerous companies to complete the construction of the facility. Each stage of construction needs to be carefully scrutinized and completed without deficiencies and be approved by senior management prior to advancing to the commissioning phase.

At this stage, operators need to have an approved preliminary decommissioning plan in place and an appropriate funding mechanism identified to ensure that decommissioning and remediation activities can be completed by the operator at any subsequent stage. This removes any burden from the government or the public in the event that the operator abandons the project at short notice.

2.2.2.7. *Mine and processing facility commissioning*

Commissioning can be defined as a series of systematic steps to ensure that all systems and components of the mine and processing facility are designed and installed as per design and they function to ensure safe and reliable operation. Ideally, initial commissioning (e.g. functional testing) of all systems and components is a specific and planned part of the construction cycle, as the contractor delivers the facility to the operator. Final commissioning with uranium ore needs to be delayed until all systems and components are determined to be compliant with design. Commissioning or startup of a mine and processing facility with mine equipment and uranium ore fed into the processing facility increases the risk of a serious safety incident (including fatality) or a significant environmental release that can impact public safety, unless it is well planned, and the proper regulatory approval is given. A formal, structured commissioning plan needs to be developed and executed so that commissioning is completed in a systematic and safe way as the mine and processing facility advance towards full production. The construction, commissioning and ramp-up of a mine and processing facility may take three to five years, depending on the complexity of the project.

At this point, the conditions outlined in Milestone 2 need to be met and the owner/operator is ready to advance to Milestone 3, namely ready to operate a uranium mine and processing facility.

2.2.2.8. *Understanding decommissioning and remediation requirements*

Good environmental site planning in Phase 3 includes the full life cycle plans, through to the post-decommissioning period, so that the project includes consideration of the end stage and ensures sustainability from cradle to grave. The operator needs to propose acceptable decommissioning and remediation plans for the orderly closure of the site even before the initial construction licence is issued. This planning provides an opportunity for the stakeholders who are engaged in the EIS phase for the first licence to also be informed of and support the final site configuration or close-out options.

The decommissioning and remediation plans need to address key factors, such as the following:

(a) Any infrastructure or access roads that will remain.
(b) Site topography, revegetation and general regrading to local standards.
(c) Mine rock piles resloped and covered as necessary.
(d) Mine areas returned to a natural configuration. Waste management sites closed and waste isolated.
(e) Environmental monitoring and surveillance after decommissioning to ensure that mine closure activities are adequate and they are functioning as planned.
(f) The options for long term institutional control. Financial guarantees to cover all costs associated with decommissioning and remediation need to be considered at the first construction licence stage and updated with every subsequent licence thereafter.

2.2.3. Milestone 3: Ready to operate a uranium mine and processing facility

At this stage, the operator is ready to begin to mine and process uranium ore, including its shipment off-site for further processing. The Member State needs to have a regulatory framework that is fully functional, with standard operating procedures and related regulatory oversight, as well as reporting frameworks to oversee the operation of the facility, including transportation safety. The regulatory body needs to ensure that the operator has an effective management system and related staff capabilities to ensure that the operation meets current regulatory requirements.

The regulatory requirements need to be the foundation for operations, either looking to increase capacity or to come on-line for the first time. Revitalized or new operations need to, at a minimum, meet current regulatory requirements for safety, environmental performance and compliance with required management systems. As technology advances, new or revitalized operations are expected to adopt the best available technology to optimize production efficiency while ensuring protection of workers, the public and the environment. In addition, prior to commissioning a new operation or revitalizing an existing operation with the intention of increasing production capacity, a detailed risk assessment on critical aspects of the uranium mine and processing facility needs to be completed, followed by the development of a risk mitigation strategy [8, 23] to ensure sustained safe and reliable production.

2.2.4. Milestone 4: Ready to decommission and remediate a uranium mine and processing facility

Prior to decommissioning and remediating a uranium mine and processing facility, a Member State needs to have regulatory infrastructure developed based on international guidance such as IAEA Safety Standards Series No. GSR Part 6, Decommissioning of Facilities [24]. A Member State that has uranium mines and processing facilities that are either reaching the end of their life or are already closed needs to ensure that the operator (or in some cases the State) meets the conditions outlined in national regulations for decommissioning and remediating closed uranium mines [25, 26]. A comprehensive closure plan including decommissioning and remediation, complete with monitoring activities, needs to be developed by the owner/operator in accordance with the regulatory requirements, noting that different stages of closure may require separate approvals from the regulatory body (or bodies).

Closure of a mine and processing facility is complex from both an operational and a regulatory perspective. To prepare for closure, the first step is to complete all mining and processing activities. The operator then needs to complete the decontamination and demolition of all required mine and processing facility infrastructure and to have a well defined plan for the management of mine waste and effluents. The final step is for the operator to remediate all affected areas to a predetermined condition suitable for final land use. All of these activities require review and approval from the relevant regulatory bodies before commencing decommissioning and remediation. Remediation may be a long term process, in some cases lasting several decades, and both the regulatory body and the operator need to be aware that remediation may require long term monitoring until a final state is confirmed. Upon completion of all decommissioning and remediation activities, the operator may apply to transfer ownership of the lease to a representative government body through a prescribed institutional programme.

The closure plan needs to be assessed and approved by the regulatory body and may include consultation periods with interested parties. The operator needs to show due diligence during decommissioning activities that is verified through ongoing monitoring. In addition, operators need to have funding and qualified personnel in place to ensure that decommissioning and remediation activities are completed, and the impacted site or area is returned to an end state agreed with relevant interested parties and approved by the regulatory body. The IAEA Safety Glossary [27] defines the end state as follows:

> "A predetermined criterion defining the point at which a specific task or process is to be considered completed."

The definition has the following clarification:

> "Used in relation to *decommissioning activities* as the final state of *decommissioning of a facility*; and used in relation to remediation as the final status of a site at the end of *activities* for *decommissioning* and/or *remediation*, including approval of the radiological and physical conditions of the site and remaining *structures*."

The operator and Member State need to be aware that decommissioning activities may take a decade or longer to complete.

Mines or processing facilities that are put into a care and maintenance state need to do so in accordance with the relevant guidance and licences issued by the regulatory body. The operator needs to present a comprehensive care and maintenance plan to the regulatory body for review and approval that demonstrates that the facility is in a safe state and that workers, the public and the environment remain protected. An overview of the general safety requirements for the protection and safety of workers and the public is given in IAEA Safety Standards Series Nos GSR Part 3, Radiation Protection and Safety of Radiation Sources: International Basic Safety Standards [28], and GSG-7, Occupational Radiation Protection [29].

In developing the care and maintenance plan, it is necessary to consider decommissioning and remediating areas of the lease that will no longer be used should the mine or processing facility resume operation in the future. This may include structures such as mine rock dumps or waste management facilities or any other disturbed areas that will no longer be required should operations resume. In addition, all activities that support environmental compliance (e.g. treating tailings pore water/supernatant or runoff from contaminated waste rock dumps) are to be sustained while an operation is in care and maintenance, in accordance with the appropriate licence issued by the regulatory body.

2.3. PRIVATE AND PUBLIC DECISIONS

The government needs to consider the roles that public (government) and private enterprise can undertake under its jurisdiction for the development of a uranium production programme. This may depend on whether national legislation defines uranium as a strategic mineral under exclusive ownership and development by the government and its agencies or by a privately owned metal resource company. Irrespective of the ownership structure, legislation and subsequent regulations need to consider radiation protection and international safeguards and security arrangements.

A government agency such as a geological survey will typically be involved in gathering and publishing general geological, geochemical and geophysical information, including maps and geological publications. This could be on a national scale or, in some larger countries, on a state or provincial government scale. In addition to general geological information, a geological survey or the geological branch of a national atomic energy agency, authority or commission may undertake targeted studies on uranium occurrence and prospecting in a country. This can be known as pre-competitive geological information.

Further exploration for uranium could then be taken up by private companies (as in Australia, Canada, Namibia, South Africa, the United States of America), a government agency or government owned company (e.g. in Brazil, Jordan) or by a combination of private and government or government owned organizations (e.g. joint ventures in Kazakhstan).

Similarly, if a potentially minable deposit is discovered, the next stage of resource delineation and staged feasibility studies could be undertaken by the government agency or government owned company (e.g. in Brazil, Jordan, Viet Nam) or by private companies (e.g. in Mauritania, Namibia, Turkey).

There are many types of private–public partnership. They can involve passive government equity in private companies, joint ventures between government owned and private companies (e.g. in Kazakhstan), partial financing from parastatal or state owned organizations, or other arrangements.

Normally, a form of monetary, infrastructural or social return is negotiated between an operator and state, provincial or national governments. This may include direct and indirect taxes, royalties, tax breaks or incentives; provision and sharing (with possible handing over) of infrastructure (e.g. water supply, roads, electricity supply), training and scholarships; and provision of, or assistance with, education and health services, and can take many other forms. This publication notes the importance of these aspects but does not attempt to analyse or provide guidance on the most appropriate forms of private–public arrangements or societal returns.

3. MILESTONES

This section provides additional details on the 16 aspects (see Table 1) associated with the development of a uranium production programme, with each of these aspects requiring specific actions during each phase of the programme. Completion of these actions corresponds to satisfying the conditions for achieving the associated milestone. The order of these aspects does not imply importance or hierarchy. All aspects are important in the development of a uranium production programme and require appropriate attention.

3.1. NATIONAL POSITION

The government needs to adopt a clear policy stating long term support for uranium exploration, mining, processing, transportation and sale of UOC and to communicate that intent locally and nationally. The government policy needs to establish measures to ensure that the operation of these facilities meets the highest standards of safety that can reasonably be achieved [30]. The national policy may define uranium as a strategic mineral under exclusive ownership (meaning development by the government and its agencies), or uranium may be considered as one of many types of privately owned mineral commodities. The national policy may also describe the economic benefits of uranium mining to both the local and national economies. These benefits can include employment opportunities (both direct and indirect) and economic value added through taxes and royalties. Examples of these economic benefits are presented in Appendix I.

In line with the national policy, the rationale for pursuing a uranium production programme within a Member State may then be either strategic (to ensure a reliable source of uranium to support domestic needs), economic (to market uranium on a global basis) or both. Strong government support, both regionally and nationally, is vital for the successful implementation of a uranium production programme as part of the front end of the nuclear fuel cycle (i.e. uranium exploration, mining and processing). The intent to support and develop such a programme must be announced at the most senior level of government. Ongoing national government support is required to ensure sustainability of the uranium mining industry [21].

If attraction of foreign investment is required to fund development of uranium mining, then government support of the industry is important to attract such investment, as investors will not develop in a country where they cannot be assured of continued beneficial ownership and operation of the uranium mine. Overall, the national policy on uranium exploration, mining, processing, decommissioning and remediation needs to be stable, transparent and aligned with other relevant and related national policies. According to IAEA Safety Standards Series No. GSR Part 1 (Rev. 1), Governmental, Legal and Regulatory Framework for Safety [31]:

> **"The government shall establish a national policy and strategy for safety, the implementation of which shall be subject to a graded approach in accordance with national circumstances and with the radiation risks**

associated with facilities and activities, to achieve the fundamental safety objective and to apply the fundamental safety principles established in the Safety Fundamentals."

The national policy also identifies the basis of national legislation and the regulatory framework for uranium mining. As part of the development of a national policy, the government needs to define how regulations and an independent regulatory body for uranium mines and processing facilities will be implemented or expanded to protect the health and safety of workers and the public, regulate nuclear safety and security, and protect the environment [32]. The term 'safety' encompasses the safety of nuclear installations, radiation safety, safety of radioactive waste management and safety in the transport of radioactive material. A number of measures can also be described in the national policy to ensure that the regulatory body is independent in its regulatory decision making. This is described in para. 12 of Ref. [33] as follows:

> "The establishment of the legal framework governing regulatory activities and their associated objectives, principles and values, including the legal basis for adequate and stable financing of regulatory activities".

3.1.1. Milestone 1: Ready to make a commitment to explore for uranium

The national policy needs to support uranium exploration as part of the development of a uranium production programme. This includes funding for a national geological survey and development of the legal and regulatory framework , including specific guidelines for land claims as well as environmental regulations that relate to uranium exploration activities. The Member State needs to define potential locations where uranium exploration activity may be acceptable and areas where it is not. For example, a Member State may not allow uranium exploration to be conducted in areas that are environmentally or culturally sensitive or densely populated. An economical uranium deposit may ultimately lead to active mining and processing. Therefore, the long term socioeconomic advantages and disadvantages for exploration areas, including public support, need to be considered prior to granting regulatory approval (e.g. licence or permit to explore for uranium). International reporting codes (such as the JORC Code) can also be adhered to by governmental exploration organizations in Member States that anticipate a need to attract foreign investors who see significant strength in adherence to known reporting standards to make reasoned and well informed investment decisions regarding the nature of a project and the associated risks.

3.1.2. Milestone 2: Ready to commit to developing a uranium mine and processing facility

Integral to the development of a uranium production programme, the national policy needs to support uranium mining and processing; otherwise, uranium exploration should not be allowed. In addition, it needs to consider the life of the uranium mine and ensure that the national policy supports uranium mining, at a minimum, for the life of the mine. Finally, the national policy should define the regulatory framework for regulation of uranium mines through the uranium production cycle and beyond (i.e. decommissioning, remediation, long term institutional control) [34]. The national policy needs to include a requirement for financial security to be paid to the government in case the operator is unable to decommission and remediate the site.

Correspondingly, at this stage the Member State needs to have an effective, independent and competent regulatory body adequately financed or budgeted to develop a regulatory process to ensure that every step of mine development, operation and waste management is completed in a safe and environmentally compliant manner. Furthermore, the Member State should develop a national security policy and strategy for uranium mines and processing facilities [35].

Within the regulatory and licensing framework, a public consultation process needs to be developed as part of the national policy for uranium mines and processing facilities, as it is important to gain and maintain the confidence and support of the general public and interested stakeholders. This is accomplished by maintaining open, transparent and timely communication and providing opportunities for interaction throughout the uranium production cycle. Other planned uses of the land post-mining need to be considered at this stage.

If the mine is to be developed domestically with the government as the operator, then the Member State's national policy needs to identify support mechanisms to ensure that it has the required expertise to advance the mine development through to production. This may include enhancing university programmes dealing with mining and mineral processing and providing support mechanisms to foster research and innovation. In addition, trained mining staff are required to ensure that the uranium is extracted safely and in compliance with all applicable regulations. The Member State needs to either develop the required expertise for mine development and operation domestically or rely on outside resources to provide that expertise.

3.1.3. Milestone 3: Ready to operate a uranium mine and processing facility

To be at a point of readiness for the final commissioning and operation of a uranium mine and processing facility, the government needs to have established the basic regulatory infrastructure to licence, regulate and safely operate the mine and processing facility according to the established laws and international best practice. At this stage, the regulatory body needs to be fully funded, staffed and trained to meet the competencies to regulate the uranium mine and processing facility. Furthermore, the regulatory process needs to clearly define the roles and responsibilities of the regulator. Finally, the regulatory body should have full regulatory and enforcement authority. Additional details on the responsibilities and functions of the government and the regulatory body can be found in GSR Part 1 (Rev. 1) [31].

Member States that have active uranium mines and processing facilities and are looking to increase capacity, either through augmenting production capacity at existing (brownfield) sites or by developing and commissioning new mines and/or processing facilities, have to evaluate each project on an individual basis. It is assumed that if a Member State has mature uranium mining and processing activities, it may already have a developed set of guidelines and regulatory licensing requirements for uranium mining and processing. However, the government needs to review these requirements and regulatory licensing conditions and update them to international best practices, if necessary, whenever an operator of a uranium mine or processing facility wants either to increase production capacity or to develop a new uranium mine and/or processing facility. Whether the objective is increased production at a brownfield operation or launching a greenfield development, there needs to be a comprehensive review of the licensee. This review is based on the project proposal provided by the operator and may include an environmental and social impact study. The scope of review needs to encompass, at a minimum, safety, radiation protection, environmental monitoring, training, decommissioning and regulatory reporting. It defines the activities that need to be taken by the operator to meet current regulatory requirements when investing in capacity expansion for existing facilities or in the development of a new mine or processing facility.

3.1.4. Milestone 4: Ready to decommission and remediate a uranium mine and processing facility

Member States need to recognize that mining is a temporary use of the land. Eventually, the mineral resources become depleted and the productive life of a mine ends. The mine sites then enter a period of formal decommissioning and remediation to remediate areas disturbed by the mining or processing activities, including the waste management and mine rock areas, to leave them in a state as defined by national regulations and associated licence conditions. Criteria for any type of mine closure are developed beforehand and updated periodically according to the intended post-closure land use to protect human and environmental health. In other words, the approach adopted should, from the very beginning, include the end state.

Future work includes continuation of monitoring and assessments of data trends and projected long term performance of remediated areas and infrastructure until the site is in the required condition to be released from formal licensing. If the site is in accordance with the decommissioning and remediation plan and achieves the predicted stability during the transition phase (post-decommissioning) monitoring period, the operator may make an application to the regulator or regulators to obtain release from further monitoring and maintenance responsibilities. The post-closure period then becomes the post-licensing phase, under a national approach to long term institutional control.

The national position needs to include language that shows national support throughout the life cycle of the uranium mine and processing facility. This includes the decommissioning and remediation phase, which may take 25–30 years to complete, depending on the complexity. It is therefore important that the national position specifies that the regulatory body remains active and funded for the life cycle of the uranium mine and processing facility to ensure that all phases in the uranium production cycle have regulatory oversight.

3.2. SAFEGUARDS

Non-nuclear-weapon States that are party to the Treaty on the Non-Proliferation of Nuclear Weapons [36] are required to conclude a comprehensive safeguards agreement (CSA) with the IAEA in accordance with INFCIRC/153 [37]. This requires the State to accept safeguards on all sources or special fissionable material within its territory, under its jurisdiction or under its control. In order to strengthen the effectiveness of the international safeguards system, many countries have a protocol in addition to the CSA, which is known as the Additional Protocol or INFCIRC/540 [38]. The CSA and the Additional Protocol contain the rights and obligations of the State and the IAEA.

The country needs to be aware of the obligations in both documents regarding mining and processing operations. To implement the provisions of these documents and facilitate cooperation with the IAEA, the Member State needs to maintain a state system of accounting for and control of nuclear material (SSAC). The SSAC needs to maintain the accounting and control of nuclear material within the State and facilitate cooperation between the country, the facility operator and the IAEA in safeguards implementation [39].

All Member States with a CSA are required to provide timely information to the IAEA regarding the import and export of any material containing uranium or thorium for nuclear purposes. States with an Additional Protocol in force also need to declare imports and exports of any material containing uranium or thorium for non-nuclear purposes meeting certain requirements. Under the Additional Protocol, a State needs to inform the IAEA of its uranium exploration projects and to declare the location, operational status and estimated annual production capacity of uranium mines, uranium concentration plants and thorium concentration plants. Additional guidance for implementing IAEA safeguards agreements is available in Ref. [40].

3.2.1. Milestone 1: Ready to make a commitment to explore for uranium

Prior to achieving Milestone 1 the country needs to ensure that its legal and regulatory framework is adequate to meet its safeguards obligations. This includes establishing laws, regulations, and an SSAC to ensure that its safeguards requirements are fully met, thereby providing timely, correct and complete declarations to the IAEA. This includes responding to requests from the IAEA, including providing support and timely access to the IAEA to locations and information necessary to perform safeguards activities. This is also important during the exploration phase, as there may be many exploration projects taking place in a Member State.

The safeguards requirements for a Member State depend on the specific safeguards agreements that the country has with the IAEA. For States with an Additional Protocol, Article 2.a(x) notes that the Member State is to inform the IAEA about its nuclear development plans for the succeeding ten year period [38]. This includes the exploration of uranium deposits, the country's plans and schedule for developing new uranium or thorium mines, and plans to extract uranium or thorium as by-products from other types of mine (for additional details, see para. 8.2 in Ref. [41]).

3.2.2. Milestone 2: Ready to commit to developing a uranium mine and processing facility

In this phase, Member States need to have well developed processes for reporting safeguards relevant to uranium mining and processing activities. This includes clearly defined roles and responsibilities for the required safeguards reporting. During this stage, the Member State needs to continue to develop its regulatory framework to ensure timely reporting of relevant mining and processing activities to the IAEA. The regulatory authority of the Member State needs to develop and communicate reporting requirements with the operators involved in uranium mining and processing to obtain the information needed for reporting to the IAEA. Setting up a coordination mechanism also facilitates an understanding of the safeguards reporting requirements, which benefits the Member State, the IAEA, mine operators and other stakeholders. At this stage, a Member State may wish to enhance the capabilities of its state regulatory authority. If requested, the IAEA can provide assistance to Member States through training, workshops and additional activities (for additional details and requesting assistance, see para. 5.5 of Ref. [41]).

3.2.3. Milestone 3: Ready to operate a uranium mine and processing facility

The stage in the nuclear fuel cycle at which full safeguards requirements specified in CSAs apply to nuclear material is defined in INFCIRC/153, para. 34(c): at this point, nuclear material has reached a composition and purity suitable for fuel fabrication or for isotopic enrichment [37]. This does not apply to material in mining or ore processing activities. However, some safeguards provisions are relevant to mining activities, and the SSAC needs to keep records about this material. According to para. 34 of INFCIRC/153, these can be summarized as follows [37]:

"The Agreement should provide that:

(a) When any material containing uranium or thorium which has not reached the stage of the nuclear fuel cycle described in sub-paragraph (c) below is directly or indirectly exported to a non-nuclear-weapon State, the State shall inform the Agency of its quantity, composition and destination, unless the material is exported for specifically non-nuclear purposes;

(b) When any material containing uranium or thorium which has not reached the stage of the nuclear fuel cycle described in sub-paragraph (c) below is imported, the State shall inform the Agency of its quantity and composition, unless the material is imported for specifically non-nuclear purposes; and

(c) When any *nuclear material* of a composition and purity suitable for fuel fabrication or for being isotopically enriched leaves the plant or the process stage in which it has been produced, or when such *nuclear material*, or any other *nuclear material* produced at a later stage in the nuclear fuel cycle, is imported into the State, the *nuclear material* shall become subject to the other safeguards procedures specified in the Agreement."

Articles 2a(v) and 2a(vi) of the Additional Protocol expand on the obligations regarding material in mining or ore processing activities in a Member State [38]. Article 2a(v) requires the SSAC to specify the location, operational status and estimated annual production capacity of uranium mines, uranium concentration plants and thorium concentration plants and their current annual production as a whole. This includes mining activities that produce uranium or thorium as a by-product [40]. The IAEA can also request the current annual production of any individual mine or concentration plant.

Article 2a(vi) addresses the State's requirement to declare information on source material that does not meet the purity and composition described in INFCIRC/153 para. 34(c), specifically information on the quantity, chemical composition, use or intended use of material exceeding certain quantities at a single location, whether it is intended for nuclear or non-nuclear use [38]. This information also needs to be provided for material in smaller quantities at different locations if the aggregate amount of material in the State exceeds the thresholds specified in the article.

Article 2a(vi) requires the State to supply the IAEA with information on the quantities, chemical composition and destination or current location of pre-34(c) material of over a certain amount exported or imported by the State for non-nuclear purposes. This includes information on exports and imports of smaller amounts of material if the total amount of material exceeds those thresholds.

Under the Additional Protocol [38], the State is also required to submit a declaration of its general plans for the next ten years relevant to the development of the nuclear fuel cycle, including research and development activities. Reference [40] provides guidance to States on how to provide additional information to implement a CSA and an Additional Protocol.

At this stage, Member States need to have a well developed IAEA safeguards reporting protocol, including the organization (or agency) responsible for completing a safeguards report and the organization (or agency) responsible for reviewing, authorizing and submitting the reports to the IAEA.

3.2.4. Milestone 4: Ready to decommission and remediate a uranium mine and processing facility

The points defined for Milestone 3 are also applicable here, with the exception that uranium will no longer be mined or processed. However, during reclamation, uranium could still be produced from water treatment, so that would have to be noted. Under the Additional Protocol [38], as part of the initial declaration to the IAEA, information needs to be provided regarding both operating and closed uranium and thorium mines and concentration plants [40].

3.3. LEGAL AND REGULATORY FRAMEWORK

A suitable legal and regulatory framework needs to be in place to support a uranium production programme. An established regulatory framework demonstrates to potential domestic and foreign investors that the government is ready to support its development.

GSR Part 1 (Rev. 1) [31] defines this expectation as follows:

> **"The government, through the legal system, shall establish and maintain a regulatory body, and shall confer on it the legal authority and provide it with the competence and the resources necessary to fulfil its statutory obligation for the regulatory control of facilities and activities."**

> **"The government shall ensure that the regulatory body is effectively independent in its safety related decision making and that it has functional separation from entities having responsibilities or interests that could unduly influence its decision making."**

The regulatory body needs to develop the following with respect to uranium production:

- — Regulatory policies on matters relating to health, safety, security and the environment;
- — Legally binding regulations;
- — A mechanism for making licencing decisions based on developed laws and regulations;
- — A compliance and enforcement programme that will ensure that licensing actions and requirements are fulfilled.

Several measures can be established to ensure that the regulatory body is independent in its regulatory decision making. As described in para. 12 of Ref. [33], these measures can be grouped as follows:

"— The establishment of the legal framework governing regulatory activities and their associated objectives, principles and values, including the legal basis for adequate and stable financing of regulatory activities;

— The establishment and implementation of clearly defined processes for regulatory decision making;
— The establishment and implementation of a clearly defined competence management programme for the regulatory body which includes an internal management programme for human resources and provides the necessary means to secure independent scientific and technical support for the regulatory activities, with international co-operation as an important component."

The legal and regulatory framework for uranium production must cover all applicable mining laws, as well as those specific to the uranium production cycle from exploration to decommissioning and remediation[1] [42]. National legislation also needs to be developed to ensure effective legal control over the export and import of UOC or other nuclear devices required for operation. Existing legislation in areas including industrial and radiation safety, human resources management (labour law), financial law, contractual law, reporting requirements for publicly traded companies (if applicable), surface lease agreements (land claims) and transport needs to be followed or modified to meet the requirements of the uranium production cycle [31]. Finally, all relevant laws need to provide clear, enabling regulations. The legal and regulatory framework for mining, including uranium mining and processing, needs to provide for fair and transparent licensing processes, royalty mechanisms and tax structures that lead to predictable outcomes [21].

The hierarchy of the regulatory framework is illustrated in Fig. 2 and addresses a broad range of Member States. Starting with national laws, the enabling legislation or act is developed (first tier) based on the constitutional powers of the Member State. The second tier illustrates the supporting regulations, which are based on the developed legislation. Any authorizations, approvals or licences will be issued in accordance with these regulations. Further requirements can be provided as licence conditions.

Regulations or decrees in accordance with the legal system of the country are issued by a government ministry or other 'competent authority', such as the regulatory body specified under the law. Whereas the law provides the general framework within which a certain activity or type of activity may take place (for instance, a law on environmental protection or a labour law), the regulations give specific explanations on how the law is to be applied in practice [43]. The

[1] A uranium mine and processing facility is separated from those for other commodities because the mineral is a naturally occurring radiological material that, when developed through a mine and processing facility, has added requirements for radiation safety, international safeguards and security arrangements.

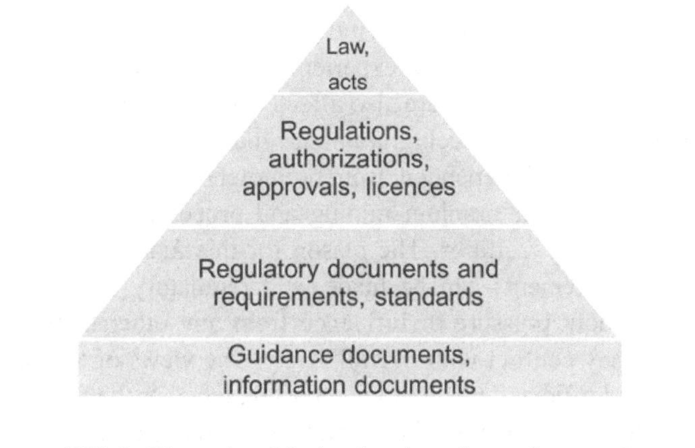

FIG. 2. Hierarchy of the legal and regulatory framework.

requirements that apply to a specific installation or to a specific activity are given in the authorization or licence that is to be granted to that installation or activity before it starts. These are known as licence conditions. The authorization is more detailed or specific and is issued in accordance with the regulations and any regulatory documents, requirements or standards so specified (third tier). Thus, the detailed authorization is written so that it applies to one facility or activity.

The principal purpose of establishing regulations is to codify requirements of general applicability. By providing well founded and clear statements of administrative and technical requirements, regulations serve to provide consistency and stability in the regulatory process.

Regulations are usually more technical than the corresponding law but are part of the national legal system. Their purpose is to achieve safety through the establishment of detailed requirements regarding the application of the law and to provide a framework for more detailed conditions and requirements to be incorporated into individual licences. To avoid misinterpretation, regulations need to be clear, unambiguous and precise [43]. However, any significant changes to the legislative framework, standards and limits should not be made without prior consultation with the operator and other affected stakeholders.

Also important is the last tier (fourth tier). The regulatory body needs to provide further supporting information or guidance on regulatory requirements on how to use the standards and on the broad use of relevant regulatory guides or information. These documents are not regulatory requirements per se, but rather clear directions and examples that may better inform the operators, the public and other stakeholders on how the regulations are best applied.

In practice, national regulations often combine performance oriented requirements with prescriptive requirements. The relative importance of these two approaches depends on national policies and strategies because some

Member States have a strongly prescriptive approach to their regulations while others do not. The knowledge and experience of the operators and the level of experience of a regulatory body are also affected by these approaches.

A regulatory body overseeing uranium production and waste management needs to be separate and independent from promoters of any nuclear technologies, resource development, or uranium mining and processing and from operators of uranium processing facilities. The reason for this independence is to ensure that regulatory judgements can be made, and regulatory enforcement actions can be taken without pressure or influence from any other interests (direct or perceived) that may conflict with overall safety. The views of the general public and of any elected official depend in large part on whether the regulatory body is considered to be independent of the organizations that it regulates. Section 2 of Ref. [33] summarizes the key features and challenges regarding independence in regulatory decision making.

The main areas of regulatory review for a uranium mine or processing facility include the following:

(a) Site characterization;
(b) Requirements for acts and regulations;
(c) Stakeholder engagement;
(d) Design principles and hazards;
(e) Construction methods, adequate controls and as-built plans;
(f) Management systems and human performance programmes;
(g) Radiation and environmental protection;
(h) Conventional safety;
(i) Water management;
(j) Waste management (e.g. mine rock, tailings);
(k) Transportation;
(l) Emergency planning;
(m) Site security;
(n) Safeguards;
(o) Decommissioning;
(p) Remediation and associated financial guarantees;
(q) Social and cultural aspects.

As stated in Ref. [33]:

"Regulatory bodies have three basic functions: (1) to develop and enact a set of appropriate, comprehensive and sound regulations; (2) to verify compliance with such regulations; and (3) in the event of a departure from licensing conditions, malpractice or wrongdoing by those persons/organizations under regulatory oversight, to enforce the established regulations by imposing the appropriate corrective measures."

The regulatory body is responsible for reviewing applications for licences based on regulatory requirements, providing recommendations to senior government officials and enforcing regulatory and licence requirements. Finally, the Member State may choose to issue separate licences that align with each stage or milestone in the uranium production cycle. Examples of licences include the following:

(a) Licence or permit to explore for uranium;
(b) Licence to construct a uranium mine and processing facility[2];
(c) Licence to operate a uranium mine and processing facility[2];
(d) Licence or permit to allow the sale and transport of uranium products nationally or internationally;
(e) Licence to decommission and remediate a uranium mine and processing facility;
(f) Release from licensing of the decommissioned and remediated uranium mine and processing facility for institutional control.

In many countries, regulatory authorities responsible for health and safety, radiation safety (e.g. nuclear substances), environmental protection, import and export, mining, etc., existed prior to the establishment of the regulatory body that is responsible for uranium mining and processing and the management of its radioactive waste. Therefore, the legislator needs to clearly allocate responsibilities among the various authorities. In particular, the powers and responsibilities of the regulatory body regulating uranium mining and processing needs to be clearly defined in the legislation.

Mechanisms to resolve jurisdictional conflicts between national authorities, or national–regional type authorities, need to be established. In such instances, a memorandum of understanding or an administrative agreement between authorities needs to be formulated to clearly define the conditions under which each authority will have the lead regulatory responsibility and coordination.

---

[2] Some forms of commissioning are associated with the end of construction. This may include rotational checks, process control checks, etc., to ensure that the equipment functions as per design and is constructed properly. Commissioning then continues during the operations stage once there is a formal sign off from construction to operation. Commissioning at the operations stage includes first commissioning the process using only water in the feed to ensure the integrity of the process prior to ramping up production using ore feed material and process reagents. The licence to construct and operate may include these two main stages of commissioning, as discussed in Section 2.2.2.6.

The regulatory bodies responsible for regulating uranium mines and processing facilities need to define how they will operate in a coordinated manner to limit regulatory gaps and contradictory overlaps and provide a defence in depth approach. They should harmonize, where practical, and avoid regulatory delays, confusion or contradictions [43].

GSR Part 1 (Rev. 1) [31] explains the need to coordinate efforts when more than one regulatory authority may be involved, as follows:

> **"Where several authorities have responsibilities for safety within the regulatory framework for safety, the government shall make provision for the effective coordination of their regulatory functions, to avoid any omissions or undue duplication and to avoid conflicting requirements being placed on authorized parties."**

At a working level, regulatory staff and compliance officers can form a working group relationship in which information and findings or issues are discussed. Such a joint regulatory group can be highly effective in cooperative approaches and regulatory defence in depth.

3.3.1. Milestone 1: Ready to make a commitment to explore for uranium

A Member State that is making a commitment to explore for uranium for the first time needs to initially review existing national legislation for conventional exploration to determine whether modifications of the existing legislation can allow uranium exploration, as noted in Section 3.2. Member States could consider input from experts in the development of legislation for the uranium production cycle and refer to other Member States that have a well developed legislative framework for uranium exploration for guidance.

The legal and regulatory framework for this milestone has to address topics such as the requirements for staking land claims, fees, taxation, mineral rights, areas identified within the country where exploration is banned (if applicable), legal aspects associated with stakeholder engagement, interaction with indigenous land claims (if applicable), exploration licence application, review and approval process, environmental protection, health and safety, including radiation safety, and transport of radioactive ore samples in public areas.

The legal and regulatory framework provides for the establishment of regulatory oversight of exploration activities by the regulatory body. Activities of the regulatory body at this stage include authorization, inspection and enforcement for exploration activities [44]. Regulatory oversight for uranium exploration falls under the control of another established regulatory body that also addresses radiation safety issues. The regulatory oversight at this stage is not as comprehensive or resource intensive as it is for uranium mining and processing, but still requires some controls for the radiation safety of workers, the public and the environment.

At this stage, the owner/operator (e.g. the exploration company) is expected to be interested in the continuity (duration) of exploration rights, exclusivity of the area staked (claimed) for exploration activities and the legislation for mineral rights that may be discovered [42].

3.3.2. Milestone 2: Ready to commit to developing a uranium mine and processing facility

Member States engaging in uranium mining for the first time need to have established the legal framework required for the uranium production cycle. At this stage, all relevant international legal instruments specific to the uranium production cycle need to be reviewed by a competent authority and all relevant legal aspects have to be incorporated into the national legal framework. In addition, a comprehensive expert review of any existing legislation in the Member State needs to be conducted to ensure that all legal elements are current, are not in conflict, and meet regional, national and international requirements for uranium production. This may include any legislation relating to the national policy for uranium mining and processing, including economic and commercial considerations. Legislation needs to be developed for specific regulation of uranium mines and processing facilities, filling gaps identified in existing legislation. A legal system for licensing (including a mining lease), inspection and enforcement for all aspects related to the uranium mining industry (e.g. radiation protection, radioactive sources, safety, security, safeguards, transportation, export and import controls, environmental law, waste management) also needs to be considered. IAEA Safety Standards Series No. GSG-13, Organization, Management and Staffing of the Regulatory Body for Safety [44], summarizes the core regulatory functions and processes that can be applied to uranium mines and processing facilities.

Legislation needs to outline the legal requirements for the operator of the uranium mine and processing facility to provide closure plans for decommissioning and remediation before construction begins. The laws for closure plans have to include legislation for financial assurances to be paid to the government so that if the mining company goes bankrupt or abandons the site, the government has access to the funds required for effective decommissioning and remediation. The legislation needs to prescribe the frequency at which these closure plans are reviewed during the operational phase of the mine and processing facility to ensure that they are current based on the status of the mine and processing facility. The closure plans have to employ industry good practice and ensure that sufficient financial assurances are available to complete decommissioning and remediation.

The legislation developed at this stage also needs to clearly define the responsibilities of all authorities involved in the uranium production cycle and cover all legal aspects associated with the uranium production cycle, which include, but are not limited to, radiation protection, safety, water and waste management, environmental law, bond, liability coverage, labour laws, safeguards, security, transport of UOC, decommissioning and remediation.

Applicable regulatory approvals need to be requested at this point, prior to advancing the mining project to the construction phase or restarting an existing mine. Once regulatory approval has been granted to construct the uranium mine and processing facility, construction may begin. As part of the construction phase, the regulator needs to be aware that commissioning activities are initially undertaken with benign material (e.g. water, clean rock), that the operator has to demonstrate that the mine and processing facility were constructed in accordance with the approved design and that all systems, structures and components operate safely and as per design intent. At this stage, a construction firm demonstrates that it has completed its contract obligations and can turn the as-constructed part of the facility over to the operator.

For Member States looking to reinvigorate uranium mining, the existing legislation may or may not align with the technology or business climate in the current uranium production cycle. Legislation may therefore have to be updated to reflect current technological, financial, legal or regulatory elements within the uranium production cycle before Phase 3 is initiated.

3.3.3. Milestone 3: Ready to operate a uranium mine and processing facility

At this stage, all required legislation and the associated regulatory framework for uranium mining and processing needs to be complete. Funding needs to be guaranteed and allocated for human resources and infrastructure requirements to ensure the sustainability of the relevant legislative processes and the independent regulatory body [31, 33, 45]. At this stage, the owner/operator needs to have the legal right to mine, process and potentially sell the UOC.

Any Member State with a long history of uranium mining and processing that considers restarting production may already have a well developed legal and regulatory framework that supports the safe, reliable production of uranium. However, the existing framework needs to be reviewed to ensure that it aligns with the proposed increase in capacity and capability. This recommendation applies to the restart of an existing uranium mine, increased production rates at an existing mine and processing facilities, as well as the development of new mines and processing facilities within the Member State. In the case of existing mines for which an increase in production capacity is contemplated, a Member State needs to review the means by which this increase will be achieved and if the existing framework provides adequate legal and regulatory oversight. If not, the legislation needs to be revised to reflect current conditions.

At this stage, the owner/operator must apply for a licence to operate the constructed uranium mine and processing facility. It is the responsibility of the owner/operator to demonstrate to the regulatory body that safety management systems and plans and programmes have been established that are appropriate to ensure safe and secure operation. In addition, completion of construction needs to be demonstrated and any deviations from the original engineering design have to be identified and described. Final commissioning and operational plans for the mine and processing facility need to be submitted as part of the application for a licence to operate. In relation to environmental and waste management, an application for a licence should also contain the effluent and environmental monitoring programmes.

The regulatory body needs to conduct a thorough review of all construction, operational, conventional safety, radiation safety and environmental aspects associated with the uranium mine and processing facility. This review needs to also include a detailed site visit with the owner/operator. If all these aspects are determined to be complete and effective for initial commissioning and operation, then the regulatory body may consider approving that activity under either an interim licence or as part of a hold point for the final phase of a construction license. This allows the owner/operator to commission the mine and processing facility, but not advance to full scale operation. All required and relevant commissioning tests and as-built reports need to be completed and demonstrated to the regulatory body prior to the issuance of the operating licence.

3.3.4. Milestone 4: Ready to decommission and remediate a uranium mine and processing facility

Mining and processing constitute a temporary use of the lands. As such, the mine life, processing facility operating life and/or capacity of the waste management area will eventually come to an end. When the end of a uranium project approaches, the closure plan, incorporating decommissioning and remediation, will need to be updated and reassessed. The regulatory framework may require the owner/operator to conduct an environmental assessment specific for decommissioning and remediation, and eventually for long term institutional control, and initiate relevant stakeholder engagement. IAEA Safety Standards Series No. WS-G-5.1, Release of Sites from Regulatory Control on Termination of Practices [46], provides a comprehensive summary of the legal and regulatory framework, as well as activities that need to be completed prior to a site being released from regulatory control. This may serve as an effective mechanism to assess the current environmental status of the facility and verify the plans for decommissioning and remediation to ensure that the agreed and approved end state for the lands and environment will be achieved.

The plans also need to include a detailed monitoring programme to evaluate progress on remediation and towards achieving the approved end state. The decommissioning and remediation environmental assessment, which includes the current decommissioning and remediation plans, has to be reviewed and approved by the regulatory body in consultation with the stakeholders and in accordance with the established legal and regulatory framework. However, the operator or responsible party needs to have a clear understanding of the operational and financial requirements that have to be met to decommission and remediate the site in preparation for long term institutional control.

Once the plans are reviewed and approved, the regulatory body has to provide frequent oversight at the operation to ensure that the decommissioning and remediation plans are being followed and that the monitoring programme, including any applicable bonds or similar financial instruments in place for funding decommissioning and remediation, has been implemented. Treatment of tailings, including supernatant and pore water, as well as runoff from waste dumps, is a critical activity during decommissioning and remediation and is of particular significance for local stakeholders such as municipal authorities or local residents. Therefore, the regulatory body needs to ensure that the infrastructure and expertise are in place to regulate this critical activity. Decommissioning, remediation and post-decommissioning monitoring activities may last up to 25 years, or longer, depending on the complexity of the mine and processing site. The Member State needs to ensure that funding is in place through a bond or similar financial instrument to support these important activities, ideally as an element of the original operational licence.

Uranium mines that are in a care and maintenance state in anticipation of reopening require a detailed care and maintenance plan that needs to be prepared and presented to the regulatory body, as it could differ from the plan submitted for licence to operate. For conventional mines, water treatment of tailings, pore water and supernatant water, as well as drainage from contaminated rock dumps, dewatering wells and accumulating mine waters, will once again be a critical activity upon reopening. A monitoring programme needs to be developed to monitor activities on-site (e.g. effluent discharge) as well as near and far field downstream receptors. A uranium mining operation may remain under these conditions for several years, so the legal and regulatory framework needs to ensure that the operator retains sufficient funds and expertise to support care and maintenance over the required duration. Finally, should an operator of a uranium mine and processing facility that have been in the care and maintenance state wish to restart mining and processing of uranium, then a comprehensive assessment of the mine and processing facility infrastructure needs to be completed to ensure that the site is in a good physical and mechanical state and is safe to operate. The operational staff may also require retraining to ensure that they are competent. These reviews may be subject to assessment and approval by the regulatory body.

The government needs to ensure that mechanisms are in place to enforce existing legislation to ensure that national, regional and, where relevant and appropriate, international guidelines for decommissioning and remediating the former uranium mine are followed. This includes a review of financial guarantees or secured funds legislation, as well as the legal requirement for the transfer of institutional control back to the state [46]. Once all legal requirements have been met for decommissioning and remediation, the owner/operator has the right to apply to the regulatory body to be discharged from all further legal, financial and regulatory obligations of the project, and the site may enter an institutional control framework. The owner/operator may also wish to apply to recover any remaining balance of the financial guarantee that was set for the project [42].

3.4. ROLES AND RESPONSIBILITIES OF THE GOVERNMENT, REGULATORY BODY AND OPERATOR

The roles and responsibilities of the government, the regulatory body and the operator change as a project advances from exploration to resource delineation, mine and processing facility engineering design, construction, commissioning, operation and ultimately decommissioning and remediation. Owing to the complexities of the uranium industry, a highly competent regulatory body and operations management team are essential to success at all stages of development. General management and financial management are required at each stage of the project; however, technical and operational management vary according to the stage.

New entrants need to consider the extent of the competence base in the government and find ways to work with operators and investors constructively but independently. When successful, such cooperation can ensure a smooth and well accepted project delivery, with objectives and decision making being mediated in a transparent manner when the government is both the regulator and (platform) investor and when government and private companies are both involved.

Government. The legal and regulatory framework for the uranium production cycle needs to establish a clear understanding of the roles and responsibilities for all organizations required to advance uranium production. The primary parties involved in uranium processing activities are the owner/operator and the regulatory bodies. Both parties have responsibilities to minimize and control impacts from the uranium production activities. The government has the responsibility to develop a legal framework (laws) that the operator needs to meet and to establish a regulatory framework, including regulations and a regulatory body with the resources to enforce applicable laws, regulations and licences. The operator has the responsibility to design, construct, operate and decommission uranium recovery operations in accordance with laws and regulations that protect the health and safety of workers, members of the public and the environment.

Regulatory body. The authority and responsibilities of the regulatory body are based on the legislation and regulations adopted or enacted by the government. To ensure that the regulations are applied correctly, it is essential to have a clear structure of roles, responsibilities and operating procedures for adequate handling of all regulatory processes. Roles and responsibilities in the oversight of, for example, a new tailings management facility and in the review of its siting or design plan need to be clearly defined among national, regional and other local regulatory bodies. Overlap of these roles and responsibilities needs to be avoided so that it is clear to the operator which body is the decision maker. It is important to set out the hierarchal structure of legal roles and responsibilities and to ensure that oversight is aligned.

One of the key roles of the regulator is to confirm (or establish, if this is not already done) the safety criteria and other regulations and guidance for the entire life cycle of the facility. The regulator is responsible for the review of new licence applications, renewals and amendment requests and for the issuance of licenses. In addition, the regulatory body is responsible for inspection and enforcement to ensure that activities are conducted in accordance with the licence and regulations. The regulatory body has responsibilities for engagement and consultation with stakeholders or interested parties.

Operator. The responsibilities of the operator of a site are defined in the laws and regulations, and include the following:

(a) Providing documentation necessary to obtain a licence or permit from the regulator, including an EIA;
(b) Constructing and operating the facility in accordance with the licence and the regulations;
(c) Protecting the health and safety of people and the environment;
(d) Providing financial surety to ensure the availability of funds for the appropriate closure of the facility;
(e) Decommissioning the facility in accordance with the licence and regulations;
(f) Providing opportunities for stakeholder engagement.

The responsibilities of the owner/operator are explained in GSR Part 1 (Rev. 1) [31] and summarized in Requirement 5 as follows:

> **"The government shall expressly assign the prime responsibility for safety to the person or organization responsible for a facility or an activity, and shall confer on the regulatory body the authority to require such persons or organizations to comply with stipulated regulatory requirements, as well as to demonstrate such compliance."**

3.4.1. Milestone 1: Ready to make a commitment to explore for uranium

At this stage, there needs to be support from national and local governments and local communities to explore for uranium and to develop the physical and regulatory infrastructure to support uranium exploration. An exploration company needs to have strong management personnel who understand the legal, regulatory, cultural, safety, environmental and social aspects associated with uranium exploration. Effective management at this stage is essential to obtain long term support for uranium mining. Exploration companies need to ethically follow the legal, cultural and social rules within the area which they are exploring and act accordingly.

3.4.2. Milestone 2: Ready to commit to developing a uranium mine and processing facility

At this stage of uranium mine development, the government needs to ensure that national laws are in place regarding uranium mining and processing. In addition, the regulatory body needs to be adequately funded and staffed. The owner/operator needs to be financially and technically responsible for the development and implementation of the uranium mine and processing facility. Further, the owner/operator has to function independently of the political and regulatory bodies in the country.

The government is responsible for the following:

(a) Collaborating with the legislative body to develop and enact the required laws to facilitate the development of uranium mining and processing;
(b) Developing an independent regulatory body for uranium mining and processing;
(c) Establishing policies for the development of a required financial guarantee (i.e. trust fund/bond) to be provided by the owner/operator to ensure financial responsibility for decommissioning and remediation;

(d) Establishing a public education campaign and stakeholder consultation programme to show support for, and oversight of, safe uranium mining and processing.

The regulatory body is responsible for the following:

(a) Recruiting and training staff as required to establish an effective regulatory structure and associated licencing and regulatory processes, including an effective compliance and enforcement programme;
(b) Establishing a structured and formal management system and associated regulations in conjunction with staff training to create a safety and quality culture in order to ensure effective licensing, regulation and oversight of uranium mines and processing facilities;
(c) Communicating the independent role of the regulatory body to internal and external stakeholders.

The operator is responsible for the following:

(a) Working with an engineering, procurement and construction management firm to design, engineer and construct the uranium mine and processing facility;
(b) Recruiting and training staff required for commissioning and operation;
(c) Establishing a formal management system to ensure a quality and safety culture in which employees feel responsible for their own safety;
(d) Establishing the required conventional safety and radiation protection programmes (including processes for reporting safety statistics and radiation exposure data to the required regulatory agencies);
(e) Establishing an asset management strategy, including a predictive and preventive maintenance programme, to ensure safe and sustained operation of the uranium mine and processing facility;
(f) Establishing an environmental protection and monitoring and reporting programme that meets regulatory requirements;
(g) Establishing and delivering public information and consultation sessions for stakeholders;
(h) Developing business relationships with suppliers required for operation;
(i) Developing a financial strategy, including an annual operating budget;
(j) Developing a working relationship with regulators and international and professional organizations.

3.4.3. Milestone 3: Ready to operate a uranium mine and processing facility

Much of the discussion for Milestone 2 (Section 3.3.2) also applies to bringing a uranium mine and processing facility into full production. Operators in a Member State aiming to reinvigorate uranium mining and processing operations need to have good understanding of the scope of work required to enhance existing capacity and capability for mining and processing uranium. Furthermore, operators need to complete a thorough analysis of the project to determine the potential environmental, safety, social and cultural impacts associated with the increase in capacity. They need to present these findings to the local and/or national regulatory bodies (as applicable in the area of interest) in a framework that meets regulatory standards for this type of application. In addition, operators have to develop and implement a comprehensive communication and consultation strategy to consult with stakeholders and obtain stakeholder support. This level of communication also needs to include a discussion of training and employment opportunities for local communities as well as business opportunities for local and national businesses. The regulatory body needs to have an in-depth knowledge of the project and be able to thoroughly review and understand the impacts that this project may have on the environment, worker health and safety and local communities and, if the application to produce is approved, ultimately issue an amended or new operating licence.

3.4.4. Milestone 4: Ready to decommission and remediate a uranium mine and processing facility

Closure of a mine and processing facility begins with completion of all mining and processing activities. Then, based on a regulatory approved decommissioning and remediation plan approved by the regulator, the operator completes decontamination and demolition of all required mine and processing facility infrastructure. The operator has to have a well defined plan for the management of mine waste and effluents. The operator then remediates all affected areas to a predetermined condition suitable for final land use. This may require long term monitoring until a final state is confirmed. Once all decommissioning and remediation activities are complete, the operator transfers ownership of the lease to a representative government body through a prescribed institutional programme.

A significant concern of local, national and international communities regarding uranium mining and processing is the long term impact on the environment, economy and cultural way of life in the local area (e.g. agriculture, hunting, fishing, recreation, community development near an abandoned or reclaimed mine or process area). Operations management needs to develop a strategic plan to conduct communication and education sessions in the community and to involve local stakeholders in the planning and decision making for decommissioning and remediation of the site and for its long term stability. Operational management then needs to develop a robust decommissioning and remediation plan and show commitment to fully remediate the site to the agreed and approved end state to ensure minimal long term environmental impact. Regulatory bodies need to have the capacity to review the decommissioning and remediation plans (with external consultants, if required) to ensure that the plans meet regulatory requirements and will achieve the agreed and approved end state, regardless of the necessity of an institutional control programme after remediation.

3.5. STAKEHOLDER ENGAGEMENT

Engaging with a variety of stakeholders is essential throughout all phases of a uranium production programme. The lack of an effective stakeholder engagement programme has been ranked consistently in the top ten business risks for mining and processing projects, and in December 2018 it was ranked as the single most important risk [47]. The uranium mining and hydrometallurgical processing industry face unique challenges with respect to stakeholder understanding and perceived risk of several factors. Some examples include potential radiological health impacts to local communities and biota during production or the impact and long term management of waste generated from such activities. Thus, effective communication and consultation are necessary to provide stakeholders an opportunity to voice their concerns, opinions and perceived risks. This dialogue allows experts to answer questions, educate and provide accurate, easy to understand information to stakeholders. Effective stakeholder engagement occurs early and often in successful projects, and consultation with stakeholders is integral to all phases of the uranium production programme. Furthermore, an independent and trusted regulator plays an important part in the stakeholder engagement process. Paragraph 1.2 in IAEA Safety Standards Series No. GSG-6, Communication and Consultation with Interested Parties by the Regulatory Body [48], states:

> "Communication and consultation are strategic instruments that support the regulatory body in performing its regulatory functions. They enable the regulatory body to make informed decisions and to develop awareness of safety among interested parties, thereby promoting safety culture. The establishment of regular communication and consultation with interested

parties will contribute to more effective communication by the regulatory body in a possible nuclear or radiological emergency."

To support effective stakeholder engagement, strong and sustained local and national government support for uranium mining and processing throughout the life cycle of a uranium mine and processing facility is necessary. The success and sustainability of any uranium mining and processing project is dependent on both government support and acceptance by a wide range of stakeholders. Each organization with responsibility in a uranium production programme — the government, the owner/operator and the regulatory body — has a role in carrying out effective stakeholder engagement activities throughout the life of facilities. These organizations coordinate outreach activities while concentrating on their distinct role to address stakeholder concerns [49]. The main objectives of stakeholder engagement are the following:

(a) To facilitate open and transparent communication;
(b) To build trust and engage with stakeholders;
(c) To provide opportunities for stakeholder consultations;
(d) To inform and educate stakeholders of potential benefits and risks;
(e) To demonstrate accountability to stakeholders.

'Stakeholder' is a broad term and for the purposes of this publication it is defined as individuals or groups who have a specific interest in a given issue or decision or in the performance of an organization. This includes the general public (particularly communities surrounding the area of the uranium mine and processing facility), local indigenous or native groups recognized by the government, employees of the mining company, groups that may undertake business with the mining company, the owners or shareholders of the company, operators, suppliers, partners, trade unions, the regulated industry or professionals, scientific bodies, governmental agencies, regulators, the media and neighbouring countries. When developing a uranium project there are two general types of stakeholders: internal and external [8, 50]. Internal stakeholders are those involved in the decision making process, while external stakeholders may be affected by the potential outcome of the project. Early involvement of both stakeholder groups is essential to achieve project goals and gain stakeholder support for uranium mining and processing [21].

General public involvement in all phases of a uranium mining and processing project is best achieved through open and transparent dialogue between the owners/operators of the project and other stakeholders [49]. Uranium mine and processing facility regulations may dictate when structured and formalized stakeholder engagement is required, for example, during the

EIA process. All concerned citizens need to be provided with access to relevant information and have the opportunity to participate in public consultations. Dialogue with key stakeholders is an important step in gaining the support required to advance uranium mining in a Member State. Moreover, public acceptance for developing and sustaining uranium mining in a Member State will depend on the competence and credibility of the organizations and individuals responsible for the mining programme. The regulatory body and owner/operator needs to be competent and open to sustain public confidence. Requirement 36 of GSR Part 1 (Rev. 1) [31]states:

> **"The regulatory body shall promote the establishment of appropriate means of informing and consulting interested parties and the public about the possible radiation risks associated with facilities and activities, and about the processes and decisions of the regulatory body."**

An important question that needs to be addressed during all phases of the uranium production life cycle is who the stakeholders are and how effective engagement will be achieved. The government, the regulatory body and the owner/operator need to be aware of their respective key stakeholder groups and the overall communication approach. Developing a stakeholder engagement strategy ensures aligned communication, which is especially critical when multiple parties are involved. This 'living document' serves as an internal playbook for the project team to continually review and refine as the project evolves and stakeholder challenges arise. Within the strategy, a stakeholder map identifies and defines the following information about key groups of stakeholders:

(a) Who they are and their location.
(b) How they receive information.
(c) How they relate to each other and the project itself.
(d) Their viewpoints or concerns regarding the project.
(e) The interests, roles or responsibilities that they represent in relation to the project; their source and type of funding.
(f) How and when communication flows to and from them to the government, regulatory bodies and operators, and through what mechanisms; how this information is managed in the best interests of effective project management and delivery.
(g) Which stakeholders need to be represented on any task force dedicated to managing the project and its wider engagement with all stakeholders.

In the government, a stakeholder engagement strategy that includes a stakeholder map may help to resolve such issues as determining which part of government is responsible for each part of the permit, regulatory and oversight process. An example of this kind of stakeholder map is given in Appendix II.

Creating a stakeholder engagement strategy and then deriving from it how best to form internal oversight committees and task forces in the government is a technique that has been applied successfully to the effective coordination and management of internal resources, and hence to interfacing and negotiating with investors and operators. Ideally, such relationships, at least during the negotiation of initial contracts and agreements, are best managed through a single point of contact in the government.

Engaging with stakeholders requires transparency and good governance, but also good leadership from the government. This starts with raising awareness and informing stakeholders as to why a given modern uranium project is in the national interest, and how it will be carried out safely prior to the construction of the mine and processing facility. During this time, open discussions on safety and risk perception need to be conducted, as there is a commonly held fear of uranium mining owing to its history of non-peaceful uses of exploitation. Additionally, many historical uranium projects had little or no controls on health, safety (e.g. radiation protection) or environmental protection of current or future generations [34].

Stakeholder engagement is a continuous discipline that evolves throughout the uranium production cycle, as priorities and stakeholder needs change. Although there are principles that can be universally applied, the application and implementation of those principles will vary depending on the organization or national context. Additional details on the development of a stakeholder engagement strategy, stakeholder mapping, and communication tools for effective, proactive engagement with stakeholders are given in Ref. [49].

3.5.1. Milestone 1: Ready to make a commitment to explore for uranium

Stakeholder engagement begins at the exploration stage of a uranium project and needs to be undertaken by the government, the regulators and the exploration company. Local stakeholders need to be advised of uranium exploration activity, not only at the start of the project, but also during and after the exploration phase. Updates need to be provided to stakeholders to inform them whether the exploration activity has identified an economical uranium resource.

It is essential to set and manage realistic expectations regarding the time required (10–20 years) from the beginning of exploration to the opening of a uranium mine and the processing of the uranium concentrate. This process needs to be carefully conducted from the moment that exploration activities begin. For example, local communities may anticipate immediate employment opportunities and wider economic benefits that, if they occur at all, may take years to materialize. Stakeholder engagement needs to be monitored throughout the life of an exploration project, as there may be a need to educate newly elected officials, new governing bodies, neighbours or businesses.

Exploration geologists, those engaged with local communities, and other stakeholders need to be trained to anticipate and have the resources available to address the hopes and fears that their presence on the ground will inevitably raise, while their sponsoring companies need to provide them with field support with respect to local community relations. If airborne surveys are planned, stakeholders need to be informed about the use of aircraft or drones. Some areas may be environmentally or culturally sensitive or densely populated, or areas may have extensive agricultural development and industrial activities with local population, and therefore uranium exploration in these areas might be incompatible with these activities. Prior to commencement of any field work, the exploration company needs to first communicate with the local administrative entities and regulatory bodies to determine what can be done in these sensitive areas.

3.5.2. Milestone 2: Ready to commit to developing a uranium mine and processing facility

The first requisite for stakeholder engagement during the development of uranium mining is for the key stakeholders to have a comprehensive understanding of the uranium project to be developed and its potential impacts and benefits through its life span. The stakeholder engagement strategy will guide this communication and outreach and will be informed by both the project team and stakeholders. From a practical perspective, this means assembling a team that is representative of the key aspects of the project, including mining and processing personnel, government officials, regulatory bodies and local community leaders. The objective at this stage is for the team to tour the proposed mining and processing operations sites and then develop a plan that identifies project milestones and goals as well as roles and responsibilities. Member States planning to mine uranium for the first time, or that have not mined uranium for a significant period of time, can involve external expert advisors to facilitate the process.

Stakeholder engagement activities and communication strategies at this phase include the following:

(a) The government communicates their support for uranium mining and processing, identifies the benefits of these activities and responds to concerns raised by stakeholders.
(b) The government communicates the development of national legislation and regulations specific to uranium mining and processing.

(c) The regulatory body describes its independent role in licensing, inspection and compliance for uranium production facilities.
(d) The regulatory body develops and communicates the formal process for public participation during the licensing process.
(e) The owner/operator of the uranium mine and processing facility describes the type of mining and processing and the way in which it will manage the safety, environmental and social aspects associated with uranium production.
(f) The economic and social benefits to local and national stakeholders are described by both the government and the owner/operator.
(g) The government, regulatory body and operator need to conduct knowledge and opinion surveys as part of their stakeholder involvement programmes.
(h) The government, regulatory body and operator need to ensure that senior staff who communicate with the public are trained.

To inform key messaging, communications and outreach as outlined in the stakeholder engagement strategy, it is important to understand how stakeholders think, what they value and what ideas or beliefs they may have that could impact the project. This type of information can be gathered by conducting stakeholder interviews or surveys. Stakeholders such as industrial suppliers, government, regulatory and public officials, environmental groups, uranium mining experts, community leaders, health professionals and other relevant agencies and parties need to be identified and interviewed to obtain their insights into the following issues:

(a) Perceptions of the uranium mining industry;
(b) Existing regional industry strengths;
(c) Workforce development challenges/opportunities;
(d) Opportunities for local involvement in the supply chain;
(e) Infrastructure needs of the area (e.g. new roads, railway, airport, water course crossings);
(f) New and emerging market opportunities;
(g) Entrepreneurial and small business support;
(h) Positive and negative impacts (real and perceived) on the uranium mining industry.

In addition, at this stage, resident and local business surveys need to be conducted to gather input from local community members. These surveys need to focus on the impacts of uranium mining and processing operations on local business and the quality of life of residents. The purpose of these surveys is to distinguish the community's real and factual issues with uranium mining from emotional and perceived issues. The outcomes of the interviews and surveys will help to shape public consultation and efforts as outlined in the stakeholder communication strategy.

The second stage in stakeholder involvement is for the government to quantitatively and qualitatively estimate and report the socioeconomic benefits of uranium mining and processing to the local and national economies. Prior to advancing to construction and ultimately operations, the government needs to clearly define the potential benefits to local communities and the national economy and address any concerns that were identified in the public survey. The owner/operator should also address concerns of stakeholders that were formally expressed during the public notification period of the EIS. The adequacy of the responses are reviewed by the regulatory body as part of its overall assessment of the EIA for the project.

Four key areas to consider in the determination of the socioeconomic impact of uranium mining and processing on the local and national economies are economic development, government services and regulation, public health and the environment, and social impacts. These four areas can be further subdivided as follows and contribute to the design and communication tools outlined in the stakeholder engagement strategy.

3.5.2.1. Economic development

The following aspects need to be considered:

(a) Direct and indirect job growth and the types of job that will be created;
(b) Local content: the number and types of job that can be filled by local workers and those likely to be filled by workers from outside the local area or country;
(c) Forecasted revenue to local businesses, including local construction companies, from spending and capital investment made directly or indirectly by the uranium mining and processing operation;
(d) Development of regional infrastructure such as roads, bridges, the power grid and cellular communication towers;
(e) Impact on local and national tax revenues;
(f) Impact on local real estate values, including potential or perceived loss of property value for properties downstream or downwind of the mine or processing facility;
(g) Direct and indirect impact on employment levels and revenue generation after the cessation of active mining and processing.

3.5.2.2. Government services and regulation

The following aspects need to be considered:

(a) Local and national government costs for the regulation and monitoring of mining, processing, tailings and waste management, decommissioning, remediation and any associated liabilities;
(b) Impact on local infrastructure and service industry;
(c) Impact on public schools, including funding and educational opportunities;
(d) Local and state government costs for contingency planning and emergency preparedness;
(e) Review of potential impact and associated costs to upstream and downstream localities, or neighbouring regions, resulting from the mining and processing operations;
(f) Potential costs of remediating any environmental damage (determination of mechanisms to hold the owner/operator financially responsible, including a bond strategy to ensure that companies retain funds for decommissioning and remediation);
(g) Potential funding or invoicing back to the mining company to offset government and regulatory costs.

3.5.2.3. Public health and the environment

The following aspects need to be considered:

(a) Potential improvements to medical care facilities, which may have a positive impact on the quality of life.
(b) Potential and forecasted impacts on the environment and quality of life, including those from catastrophic environmental consequences (e.g. tailings dam failure). This also includes localized impacts on natural landscapes, scenic appeal, recreation and tourism, including wildlife and hunting, fishing, boating and places of historical interest that might be affected.
(c) Post-closure (e.g. decommissioning and remediation) procedures to ensure that public health and safety requirements are met and the environment is returned to an acceptable long term state.

3.5.2.4. Social impacts

The following aspects need to be considered:

(a) Effects of uranium mining and processing on the internal and external image and reputation of the region — for example, belief that the area will remain a safe place to live, work and invest;
(b) Public confidence in the company to prevent adverse effects and in the ability of the government to properly regulate such effects;
(c) Impacts on schools and private institutions;
(d) Direct and indirect employment opportunities;
(e) Benefits to local and national economies in terms of taxes and royalties;
(f) Impact on the aesthetics of the area.

An external stakeholder report needs to be prepared by the government on a regular basis (e.g. annual) that covers these points and presented to stakeholders — for example, during a public meeting — in a manner that is clear and concise and using language that can be easily understood by everyone. A translator may need to be present if multiple languages are spoken, and the materials also should be accessible to stakeholders in their native language. The public meeting needs to be conducted in a manner that allows for a facilitated question and answer session to engage with stakeholders and address any concerns. In addition, to maintain stakeholder engagement and presence with the stakeholders, the operator needs to hire or delegate an appropriately trained and experienced employee to work as a senior spokesperson to interact with stakeholders and provide ongoing updates (at intervals appropriate to the milestone, considering other influencing factors) as part of the stakeholder engagement strategy. This level of engagement shows competence of the operator and demonstrates accountability to the community. Completing these initial actions can result in a comprehensive public information and education programme that helps to gain and sustain the confidence of the local and national communities.

3.5.3. Milestone 3: Ready to operate a uranium mine and processing facility

The activities undertaken at Milestone 2 (Section 3.4.2) regarding stakeholder engagement can also be conducted during the commissioning and operation of a new uranium mine and processing facility. The operator needs to review the best practices noted in Section 3.5.2 and conduct a gap analysis to continually update and refine the stakeholder engagement strategy.

Some considerations for stakeholder engagement in this phase include the following:

(a) The government, regulator and owner/operator continue to conduct stakeholder surveys.
(b) The government continues to show support for uranium mining and processing, including the expected benefits, and responds to concerns raised by stakeholders.
(c) The regulatory body continues to engage with stakeholders and continues to provide information to stakeholders on their role, the licensing process and the inspection and enforcement programmes.
(d) The regulatory body arranges for public involvement during the licensing process.
(e) The owner/operator provides regular updates on the construction process and its preparation for commissioning and operation.
(f) The government, the regulatory body and the owner/operator inform stakeholders of on-site and off-site emergency response plans.
(g) The government, the regulatory body and the owner/operator inform stakeholders of the mechanism for ongoing stakeholder engagement as the mine and processing facility advance into full operation.

As the facility moves into the operational stage, the operator or government can ask a nearby community to initiate an independent environmental monitoring programme. Community members have to be trained on proper sampling techniques and an accredited laboratory needs to be used for sample analysis. This allows community members to periodically conduct their own independent environmental monitoring programme to verify that the facility is operating as planned. The community data could then be compared with the company and regulatory data and shared with stakeholders in an understandable format as a means to show that the site is remaining in compliance. Monitoring sites are chosen off-site to ensure that there are no disruptions or safety issues with the operator. The regulatory body needs to also consider collecting independent environmental monitoring samples at the uranium mine and processing facility site (near field) and downstream or downwind of the facility (far field). This enhances public confidence in the regulatory body, serves as an opportunity to engage and build capacity with local stakeholders and, finally, provides an independent evaluation of the environmental performance of the facility.

When an operator wants to increase the capacity of an existing operation (e.g. new open pit mine), all changes that will occur and their impact on the relevant stakeholders need to be considered. The operator will also have to work with the regulator to determine if the increase in capacity requires a new

EIA prior to amending the licence to operate or if an amendment to the existing operating licence is sufficient. In both scenarios, the operator needs to involve the stakeholders and inform them of the capacity increase and the impacts that this will have. The process described in Section 3.5.2 is also applicable in this situation.

3.5.4. Milestone 4: Ready to decommission and remediate a uranium mine and processing facility

As is the case for the preceding milestones, stakeholder involvement continues to be important for a uranium project at the end of its life. This includes facilities approaching decommissioning, those advancing towards active remediation, those already in active remediation and those being prepared for care and maintenance with the intention of reopening in the future. Operators of uranium mines and processing facilities approaching decommissioning and remediation need to keep the relevant stakeholders informed with regard to planning, decommissioning and remediation activities, monitoring programmes and the final intended state of the site (e.g. use for recreational, residential, commercial or agricultural activities under institutional control) [50]. Paragraph 2.53 of IAEA Safety Standards Series No. GSG-15, Remediation Strategy and Process for Areas Affected by Past Activities or Events [51], states: "Interested parties should have a role in contributing knowledge and information to the remediation process."

The role of interested parties (e.g. stakeholders), such as members of the public, the responsible party, the regulatory body and other authorities involved in the remediation, is to exchange information in an ongoing dialogue to help ensure that well informed decisions are being made. Representatives of interested parties need to discuss their positions, expectations and views regarding the remediation. This will facilitate the development of a mutual understanding and meaningful involvement in the decision making process regarding the planning and implementation of remedial actions.

Operators of uranium mines and processing facilities that are in care and maintenance need to keep stakeholders informed of ongoing site activities, including progressive reclamation, environmental management and monitoring, and the economic and social impacts to the relevant stakeholders. For both scenarios presented in this milestone, the best practices identified in Sections 3.5.2 and 3.5.3 need to be applied.

3.6. SAFETY AND RADIATION PROTECTION (INCLUDING EMERGENCY PLANNING) OF WORKERS AND THE PUBLIC

Laws, regulations, emergency planning and monitoring programmes are necessary to ensure the safety of workers and the general public. A Member State considering uranium mining and processing is expected to have a legal and regulatory framework for conventional safety and radiation protection in place that complies with international standards and national guidelines. The legal and regulatory framework needs to encompass all current activities, practices and facilities in that Member State. The regulatory body needs to develop an understanding of the hazards presented (biological, chemical, physical and radiological) in uranium mines and processing facilities and use that information to develop guiding regulatory principles to ensure the safety of workers and the general public. In most mines, physical safety hazards are by far the most significant. This is also true for most uranium mines, except for some with very high grade deposits where the radioactivity levels are naturally so high that only remote mining by operator controlled or autonomous equipment is possible.

Safe production needs to be a key foundation in the uranium mining and processing industry and all facilities need to strive for zero harm to their employees. The operator needs to have a sustained focus on safety to develop and maintain a strong culture for safety. The IAEA Safety Glossary [27] defines culture for safety as follows: "The assembly of characteristics and attitudes in organizations and individuals which establishes that, as an overriding priority, protection and safety issues receive the attention warranted by their significance."

To achieve this, a focused effort by the operator is needed to ensure that the core components of an effective safety programme are developed. Some of the key components of such a programme include the following:

(a) Development and maintenance of an effective and practical health and safety plan;
(b) Employment and empowerment of skilled/professional health and safety personnel;
(c) Sustained enforcement of health and safety standards and procedures, including audits;
(d) Effective training of all people on the site (i.e. knowledge of the job, hazard identification, management of risk tolerance);
(e) Training and continual development of supervisory staff with regard to safety;
(f) Development of an effective incident management system;
(g) Continual improvement in a non-confrontational environment.

Applicable radiation safety requirements will vary depending on the stage of a Member State in the uranium mining and processing life cycle. As uranium and its associated decay products are mined and concentrated, the associated requirements for radiation safety increase. Considerations for radiation safety for each generalized situation of the uranium mining and processing life cycle are provided below.

The lead IAEA radiation protection requirements publication is GSR Part 3 [28], which is generally referred to as the Basic Safety Standards. GSR Part 3 applies to all facilities and all activities that give rise to radiation risks and lays down a consistent and harmonized system for the protection of people and the environment. Additional guidance and information on safety and radiation protection are available in IAEA Safety Standards Series Nos SGS-7 [29], SSG-31, Monitoring and Surveillance of Radioctive Waste Disposal Facilities [52], RS-G-1.6, Occupational Radiation Protection in the Mining and Processing of Raw Materials [53], and in Refs [54–56].

3.6.1. Milestone 1: Ready to make a commitment to explore for uranium

The regulatory body needs to review and implement international safety standards to oversee the development and operation of an exploration project. From a health, safety and environmental (HSE) perspective, an exploration project needs to follow all national and local regulations and standards for protecting workers, the public and the environment from harm. In the absence of such regulations and standards, or to complement and reinforce them, the exploration company or mine operator undertaking exploration activities needs to develop an appropriate worker training programme, supported by relevant standard operating procedures. In line with industry best practices and norms, standard operating procedures need to address specific hazards of a task and describe measures to mitigate those risks. Standard operating procedures may be developed to address proper personal protective equipment, working in potentially remote and harsh conditions, work permits as needed, lock-in and lock-out requirements, how to conduct daily checks to ensure safe operation of equipment, a process for conducting field level risk analysis, brief safety meetings at start of each shift and an incident reporting system.

The exploration company's radiation protection programme needs to include measures to ensure that an employee's radiation exposure remains as low as reasonably achievable (ALARA), economic and social factors being taken into account. Some aspects of ALARA include the following:

(a) Engineering controls;
(b) Administrative controls;

(c) Contamination/zone control;
(d) Use of personal protective equipment;
(e) Radiation monitoring and record keeping;
(f) Implementation of good hygiene practices;
(g) Employee training with regard to safe practices to minimize radiation exposure.

Engineering controls need to take precedence before administrative controls and the use of personal protective equipment. An example of this is the installation of engineered ventilation in the uranium mine and processing facility to significantly reduce or eliminate the need for personal respiratory protection.

A training programme and the related standard operating procedures need to be developed to protect workers from potential exposure to radiation during exploration activities. Such exposure can originate from the following:

(a) External radiation exposure (beta and gamma radiation) from the handling of drill core material or during trenching activity. This is directly related to the concentration of uranium in the core material and the duration of exposure. Time, distance and, if applicable, shielding are the most effective controls to reduce exposure to beta and gamma radiation.
(b) External radiation exposure from radioactive sources used during exploration, in particular during drill hole probing.
(c) Exposure to radon progeny (alpha radiation) from drill cores and samples stored in an enclosed, non-ventilated area. Core shacks and geology workstations that are used to store or analyse uranium-bearing drill cores require proper ventilation. Drilling and related exploration activities in abandoned underground mine workings can release significant amounts of radon gas, resulting in hazardous conditions. Adequate ventilation needs to be ensured prior to drilling/exploration.
(d) Long lived radioactive dust particles that originate from splitting or crushing drill core material. Exploration workers need to ensure proper dust control (wet cutting or dust hoods) and personal hygiene measures, including the use of personal protective equipment and washing facilities for skin and clothing.

3.6.2. Milestone 2: Ready to commit to developing a uranium mine and processing facility

Development of a uranium mine (e.g. open pit, underground, in situ recovery, heap leach) and processing facility is a complex task, and safety and radiation protection aspects need to be considered as the project advances through its various stages. During the active construction phase and during handover of the completed facilities to the operators, there is a considerable amount of activity and a large contingent of personnel coming to or leaving the site. This includes contractors, subcontractors, supervisors, design engineers, inspectors and specialist trades (i.e. shaft sinking, process tanks). Worker and visitor safety and radiation protection becomes a complicated but essential service during the construction phase and cannot be overstated. Construction and underground mine development are high risk occupations and require appropriate oversight by the operator and the regulatory body. This includes detailed design of mine and processing facility ventilation, as well as other engineered controls, to maintain radiation exposure ALARA and ensure the structural safety of any constructed or mine workings. As noted, engineered controls need to be the first line of defence to protect workers from radiation exposure or safety issues. Use of administrative controls and personal protective equipment are secondary.

At this stage, it is necessary to have all conventional safety, radiation monitoring and protection programmes developed and implemented for construction, prior to the start of mining or processing of uranium ore. Reference [57] provides comprehensive and practical information on radiation protection, monitoring and dose assessments for uranium mining and processing facilities. In developing the mine and processing facility and before production starts, the radiation protection conditions that need to be met include the following:

(a) Adequate ventilation, complete with redundant systems, for underground mines as well as processing facilities to ensure that workers are protected from radon gas and long lived radioactive dusts. Similarly, for open pit mining, the operator needs to ensure that effective air filtration equipment is provided for all heavy duty mobile equipment to protect workers.
(b) A management system to set guidelines for radiation zone control (e.g. wash stations, dedicated eating/drinking areas) to protect the workers and minimize radioactive contamination of non-working areas. Smoking can only be allowed in designated areas, both from a conventional safety and a radiation protection perspective.
(c) Radiation monitoring equipment installed and operational at the mine, at the processing facility and off-site to measure background values in general and monitor air quality for potential radioactive dust migration (as part of a fully implemented environmental monitoring programme).
(d) A functioning off-site radiation monitoring programme.
(e) Radiation dosimetry requirements in place for all workers.
(f) Programmes developed to reduce radiation exposure during operation and maintenance of the mine and processing facility (ALARA programmes).

(g) Waste management practices for the management of low level radioactive waste.

The radiation protection and safety programmes for the construction or operation of a uranium mine and processing facility are more comprehensive than the programme for uranium exploration. During the development of uranium mines and processing facilities, it is recommended that the company employ a qualified radiation safety officer to develop and ultimately manage the site radiation protection programme. Aspects of this role include, but are not limited to: (i) developing relevant radiation protection training programmes for site visitors, contractors, employees, supervisors and management; (ii) recommending and implementing relevant radiation monitoring equipment to ensure that engineered controls are effective and that personal radiation exposure can be calculated; and (iii) developing a database to track and report employee radiation exposures to the relevant agencies at the prescribed frequency. The radiation safety officer is invaluable during the construction phase of the uranium mine and processing facility to ensure that radiation protection issues are appropriately addressed so that workers remain protected.

At this stage, programmes for the protection of workers from a conventional safety perspective during operation of the uranium mine and processing facility need to be developed. An effective conventional safety programme includes the following activities:

(a) Ensuring that roles and responsibilities with respect to safety are developed and understood;
(b) Conducting effective 'toolbox meetings' every day prior to the start of a work shift (review planned jobs, identify and mitigate safety risks);
(c) Continual management presence at toolbox meetings and in the workplace;
(d) Supporting the use of all staff to complete job task observations in the field;
(e) Making use of safety tools, such as job hazard analyses, to reduce safety risk for non-routine tasks;
(f) Completing independent audits of health and safety practices and procedures;
(g) Enforcing disciplinary measures for deliberate violation of safety rules, which may extend to removing personnel from the site who violate important lifesaving health and safety procedures;
(h) Developing a programme to define how safety performance at the uranium mine and processing facility will be tracked, reported and communicated to all relevant stakeholders.

At this stage the operator needs to complete a risk assessment of potential radiation and conventional safety incidents affecting both workers and the public and to develop mitigative strategies. Contingency plans need to also be developed to identify and develop a response plan for radiation and conventional safety emergencies.

Finally, at this stage the regulatory body needs to review and approve the operator's radiation protection and monitoring programmes, as well as the reporting frequency, for radiological doses to ensure that they align with regulatory requirements. The regulator also needs to develop conventional safety and radiation protection regulations and enforcement mechanisms to ensure compliance with these regulations.

3.6.3. Milestone 3: Ready to operate a uranium mine and processing facility

The aspects covered in Section 3.6.2 apply to a uranium mine and processing facility that is ready for final commissioning and moving into the operational phase. The lessons learned on safety and radiation protection oversight from the construction phase need to be addressed and implemented in the operational phase. In fact, these are continual learning and improvement programme areas.

Regulatory systems and operational management programmes need to be in place at this stage to address safety in a proactive manner. Both the operator and regulatory bodies are responsible to promote a strong safety culture. The IAEA Safety Fundamentals and Safety Standards series provide the reference for international good practices for both the regulator and the operator. Safety needs to be an intrinsic consideration for all activities associated with uranium mining and processing to ensure a strong safety culture and ultimately strong safety performance.

The following list summarizes important requirements to maintain a strong safety performance in a uranium mine and processing facility:

(a) The operator needs to work to create a strong safety culture through effective safety training and promoting positive attitudes to safety.
(b) The operator needs to accept the primary responsibility to ensure safety of the workers and the public.
(c) An effective management system is developed by the operator that provides practical guidance in areas of safety and to ensure that sufficient funding is in place to sustain strong safety performance. The management system for safety needs to be evaluated at a prescribed frequency to ensure relevance and look for opportunities for improvement.
(d) A comprehensive asset management strategy needs to be developed by the operator to ensure that mining and processing equipment and infrastructure are well maintained.

(e) Operations staff need to be effectively trained on all technical aspects, including the operation of equipment, to ensure safe and efficient uranium mining and processing.
(f) The operator needs to share experience with similar industries to understand lessons learned from safety related incidents.
(g) The regulatory body needs to be competent, independent and empowered to enforce compliance with all regulations, including safety regulations.
(h) Emergency preparedness and contingency plans need to be well developed and frequently reviewed to ensure completeness and to confirm that measures are in place to effectively deal with emergencies. This needs to include both proactive and reactive measures.

A Member State wishing to reinvigorate uranium mining and processing or to enhance its existing capacity and capability needs to review historical baseline and worker radiation exposures. This may include an audit or professional review of its radiation protection and safety programmes to ensure that they remain effective to meet the requirements of future production, including stricter regulatory requirements or best practices that may have been introduced since earlier operations ceased.

As part of a capacity increase study, the operator needs to consider radiation modelling of the enhanced mining and processing process, in conjunction with predicted future radiation exposures to employees. A competent radiation safety officer or health physicist may be required to develop the model, interpret the data and recommend mitigation measures to either reduce radiation exposures to workers or, at a minimum, maintain the radiation exposure of the workers at historic levels and below regulatory limits. In addition, the assessment needs to show that the increase in capacity will not result in higher radiation exposure to contractors and the general public working and living within the area of the uranium mine or processing facility.

3.6.4. Milestone 4: Ready to decommission and remediate a uranium mine and processing facility

The aspects covered in Sections 3.5.2 and 3.5.3, including emergency planning, also apply to the decommissioning and remediation phases. As for Milestone 2, it is likely that a number of contractors will return to the site for specialized decommissioning, demolition and decontamination activities. More equipment may be sent to the site and be subject to clearance procedures and approvals, and more materials or salvaged equipment may be removed from the site during this phase. Contamination and clearance controls are priorities to protect workers and the public off-site. In terms of worker safety, dismantling and destruction or demolition of large buildings, vessels, etc., carry increased safety risks to the workers in the area. As noted in Section 3.5.2, a conventional safety programme needs to be updated for this new phase of activities. Additional details on decommissioning of uranium mines and processing facilities are provided in GSR Part 6 [24].

Uranium mines and processing facilities that are closed or in a state of care and maintenance (i.e. not yet decommissioned or remediated) need to have a detailed radiation monitoring programme to ensure that workers and the public are not unreasonably exposed to radiation. Potential radiological hazards include increased gamma exposure from process vessels and piping that have not been properly cleaned (e.g. excessive scale buildup) prior to being put in a state of care and maintenance or decommissioning. In addition, airborne dust (e.g. long lived radioactive dust) from mine or processing facility workings (e.g. ore pads, mine workings, tailings facilities, waste stockpiles, uncleaned process equipment containing dried slurries) needs to be properly managed to ensure that workers and the general public are not exposed to dust, airborne radioactive dust or alpha emitters. Finally, uranium mines and processing facilities may have sealed nuclear sources (e.g. nuclear density gauges) that are no longer required. These sealed sources need to be properly handled and disposed of and the disposal process needs to be in keeping with international standards for disposal [58]. An international outlook and guidance on the safety requirements for the disposal of radioactive waste are provided in section 3 of Ref. [59].

3.7. ENVIRONMENTAL PROTECTION

3.7.1. General

An appropriate national and regional (when applicable) regulatory framework for environmental protection, which is based on international good practices, needs to be in place to cover all aspects of the uranium production cycle [60]. Environmental regulations for each phase of the uranium production cycle need to be well developed and comprehensive. The short and long term environmental impacts need to be based on scientific evaluation prior to initiating mining, processing or decommissioning/remediation, and effective environmental regulations need to be in place to mitigate risk and minimize the short and long term impacts of uranium production. Mining and processing are a temporary use of the land, so it is important to control the size and duration of any potential environmental impacts.

Environmental protection needs to be a key focus area through all stages of the uranium production life cycle, from exploration to decommissioning and remediation. Aspects to be considered include water resources (ground and surface), air quality (e.g. dust, noxious gases, radiation), noise, biota, amenity and wildlife (especially rare and protected wildlife). The three guiding principles in the environmental management of responsible uranium mining are as follows:

(a) Sustainable development principles;
(b) ALARA principles;
(c) Precautionary principles [8].

Sustainability of the uranium mining industry is based on a balance between environmental, social and economic requirements within a regime of strong governance. Good corporate and regulatory governance are required to ensure clear direction on the appropriate balance of these three principles. Environmental impacts need to be kept ALARA and controls have to be based on best available and practical technology. However, social and economic factors need to be taken into account when developing and implementing controls. The precautionary principle requires that effective environmental management needs to anticipate, prevent and mitigate the causes of environmental degradation.

The application of good practice ought to be implemented at all phases of the uranium production cycle (exploration, conceptual design, feasibility studies, construction, operation, decommissioning, remediation and closure). The main elements of the best practices in environmental management are as follows [8]:

(a) Baseline data collection, which includes socioeconomic and environmental characterization;
(b) Public and stakeholder involvement;
(c) Impact assessment and mitigation strategies;
(d) Design and implementation of an environmental management system and a monitoring and reporting programme;
(e) A waste management strategy that includes identification of waste streams, volumes and appropriate storage, treatment and disposal options (see Section 3.15);
(f) Decommissioning, remediation and closure plans considered prior to the development of the mine and processing facility.

Environmental planning and monitoring throughout the life cycle of the mine ensure that the expected performance is achieved through to the post-decommissioning period, minimizing the environmental effects to acceptable standards and avoiding impacts on local populations.

3.7.2. Milestone 1: Ready to make a commitment to explore for uranium

The regulatory framework within the country, region or territory (when applicable) needs to stipulate the responsibilities for regulating and monitoring exploration activities within its jurisdiction and for informing the public about them. Initial uranium exploration, which includes airborne and ground surveys, is non-intrusive and poses a low risk to public health and the environment. If a suitable target is found, then the next stage of exploration may involve construction of temporary access roads, drilling, trenching and test pitting. These exploration activities have the potential for some localized environmental impact, in particular on surface and groundwater (e.g. cross-contamination of water between aquifers), so effective regulations and regulatory oversight need to be in place. The licence/permit to explore needs to mandate appropriate conditions for exploration drilling, including management of radioactive materials and radioactive and non-radioactive waste. In addition, the licence/permit to explore needs to contain conditions that the exploration site is remediated back to the pre-existing or background conditions if no further activity is planned. During exploration, drill fluids and contaminated water need to be properly managed and any resulting radioactive or hazardous solids need to be appropriately disposed of.

Exploration companies or national geological surveys that have identified exploration areas that show promise for more detailed exploration and potential further mine development need to collect initial environmental baseline data at the exploration site. This includes basic information on soil and vegetation types and an understanding of the regional biota, geological and climatic conditions. Industry good practice has shown that environmental baseline data need to be collected for at least three years through a variety of seasonal and climatic conditions prior to commencing construction and operations [21]. This will support the environmental assessment that will be required as part of the licence to construct a uranium mine and processing facility, should this proceed after the exploration activities have defined an economical resource. Environmental baseline study guidelines need to follow regulatory guidelines and need to include, as a minimum, information on hydrological and hydrogeological conditions, flora and fauna, wildlife, biota, archaeological and heritage surveys, anthropological surveys and climate, as well as soil, water and air analysis.

3.7.3. Milestone 2: Ready to commit to developing a uranium mine and processing facility

An environmental assessment [8–10] needs to be completed by the operator on the basis of agreed and approved guidelines. This needs to include an assessment of both the environmental background/baseline conditions at the proposed uranium mine or processing facility site and of the impacts that mining or processing will have on the local and regional biota (i.e. air, land and water). An effective monitoring programme needs to be developed as part of the EIS to track environmental performance. The data collected from the monitoring programme can be used to compare the impact of the site against the baseline data collected during the EIS. The environmental assessment also needs to identify action levels, so the operator or regulator can intervene to correct any potential or emerging environmental impacts before they become serious.

The main environmental aspects and potential for long term liability from uranium mines are contaminated waste rock, ore stockpiles, mine water, surface water and groundwater. Correspondingly, the main environmental aspects for uranium processing facilities are tailings and waste management operations and the management of process and tailings water and of their impact on environmental receptors, both surface and subsurface. As such, industry good practice needs to be applied in developing operational strategies and processes to manage tailings and water management facilities within the context of the licence or permit for operating the site [61, 62]. Emissions from processing facilities (e.g. ammonia, sulphur dioxide, solvent extraction reagent, uranium/calciner dusts) need to be considered as well. Effective regulatory processes need to be developed that utilize industry good practice for regulatory oversight of these aspects.

The operator needs to indicate the industry good practices being applied with regard to environmental management to minimize the environmental impact. The proposed design of the mine or processing facility, including relevant environmental controls, needs to be included in the environmental assessment and the environmental impacts of the life cycle. In turn, the outcome of the environmental assessment process, which identifies critical or vulnerable receptors (e.g. groundwater, wildlife, airborne emissions), feeds back into the final design of the facility requiring regulatory approval. Some specific aspects that need to be considered in the environmental assessment, which make it an important planning tool, include the following:

(a) Pathways for effluent loading and impact on downstream (near and far field) environmental receptors;
(b) Impact of air emissions and dust from mining, contaminated waste rock piles, tailings facilities and processing of uranium;
(c) Identification (presence/absence), abundance and particular sensitivities of plant and animal life, and associated impact of mining and processing uranium;
(d) Impact of uranium mining and processing on local populations (e.g. via impact on groundwater, surface water, soils and food sources);

(e) Volume and origin of water used for mining and processing, or clean waters diverted before they become contaminated (e.g. open pit dewatering);
(f) Waste management strategies (e.g. tailings, mine rock segregation, development waste, radioactive slimes and sludge, radioactive and non-radioactive waste, putrescible landfill management).

These aspects need to be well understood and documented to form a strong scientific baseline and framework to monitor against future potential impacts. The operator needs to show that the proposed management strategy is effective at ensuring minimal impact on the environment during design, construction, commissioning, operation, decommissioning and remediation. This needs to be demonstrated through management actions and an effective environmental monitoring programme. Overall, these foundational management strategies need to be in place before the regulatory body can issue a licence to construct and a licence to operate.

The regulatory body needs to put in place science and evidence based environmental guidelines and discharge limits for environmental contaminants. These limits will be set with consideration to the natural background, part of which is understood by the studies completed by the operator as part of their EIA, thereby highlighting the importance of appropriate temporal and spatial studies of the receiving environment. This also includes regulations for effluent and air emissions and related guidelines and standards for the management of tailings, mine rock, radioactive contaminated waste (e.g. pipes, rags) and non-radioactive waste. These regulations and standards need to be in keeping with international standards for environmental protection and based on the best available and practical technology.

3.7.4. Milestone 3: Ready to operate a uranium mine and processing facility

At this stage, the regulatory body needs to be well developed and include regulations and guidance specific to environmental management, monitoring and reporting, including reporting and follow-up requirements for accident conditions, such as uncontrolled releases (spills). Good practice with regard to stakeholder engagement and sustainability is that regulatory requirements for environmental reporting from uranium mines and processing need to be public documents. The regulatory body needs to be fully staffed with qualified personnel who can review the environmental performance of the mine and processing facility and have the authority to enforce the established regulations.

The owner/operator needs to have a comprehensive environmental management programme that is compliant with the established regulations and aligns with the aspects identified in the environmental assessment and the operating licence. This includes both operational and statutory environmental monitoring and reporting. Furthermore, the mine and processing facility needs to have a fully staffed and dedicated environmental management team. During operation, the owner/operator needs to assess the environmental performance of the uranium mine and processing facility and pursue continual improvement opportunities based on best available technology to reduce environmental risk and impact. Action levels or triggers need to also be identified, and a corrective action programme needs to be put in place to take remedial actions before significant harm or impacts arise.

Contaminated waters collected and generated throughout the uranium mining and processing site need to be treated to ensure efficient removal of radionuclides and unwanted metals prior to releasing effluent from the site to the environment. Contaminated waters are primarily sourced from hydrometallurgical processes (e.g. raffinate from solvent extraction), from dewatering activities (e.g. mine dewatering, seepage and runoff from surface sources, including waste rock piles and ore stockpiles) and from tailings management facilities. Prior to releasing effluent from the site to the environment, the quality of the water in the monitoring ponds needs to be confirmed through sampling and analysis. Effluent should be released to the environment only after the results indicate that the quality of the water meets the requirements of authorized limits for release.

The regulatory framework needs to define the minimal frequency of site inspections and environmental programme audits by the regulatory body to ensure compliance with the conditions prescribed in the operating licence. The regulatory body needs to conduct an independent monitoring programme, including collection of treated effluent and discharge samples from the mine or processing facility, and have them independently analysed to ensure compliance with regulatory guidelines and limits.

All of these aspects for environmental management, including performance criteria (e.g. effluent release guidelines), monitoring and reporting, need to be included in the licence to operate that is issued to the operator of the mine and processing facility once regulatory conditions have been met. All lessons learned and as-built information from the construction and commissioning activities will form part of the operational environmental management programme.

Member States seeking to reinvigorate uranium mining or increase capacity may have environmental regulations in place owing to their history of mining and processing. This regulatory framework needs to be reviewed to ensure that it includes provisions and licensing guidelines for new mines and processing facilities, as well as for existing operators wishing to increase capacity and capability. If a Member State wants to approve a new mine or processing facility for a new licensee, then the steps outlined in Section 3.6.3 apply. If an operator wants to increase the capacity and capability of an existing operation, an environmental assessment may need to be completed to determine the environmental impact for increased production rates. This need is determined by the requirements of the regulatory framework. Impacts to land, water, air and their receptors have to be assessed and compared with both the original EIA and historical performance. Key environmental aspects that need to be considered when increasing production capacity and capability include the following:

(a) Impact on effluent loadings to near and far field environmental receptors;
(b) Impact on the tailings volume and on the performance and capacity of the tailings facility (geotechnical and geochemical);
(c) Mass of mine rock that will be produced and management strategy to ensure minimal environmental impact from mine rock (e.g. segregation, proper storage, geochemical controls to prevent leaching of contaminants);
(d) Impact of additional radon and long lived radioactive dust on air;
(e) Cumulative impact on the current decommissioning and remediation programme and the objective of meeting a long term institutional control outcome.

The goal of an environmental assessment at this stage is to determine whether the operation will have a higher environmental impact with an increase in production capacity or capability. If the assessment shows a statistically significant increase in the environmental impact on land, air, water or their receptors, then the operator needs to be challenged to ensure that all reasonable actions have been taken to mitigate the environmental impact. This could include the implementation of new technologies or different operating strategies (e.g. recycling of effluent back to the process rather than using additional fresh water) to minimize environmental impact.

Once all the terms and conditions of the EIA have been met, assessed and approved by the regulator, an amended operating licence can be awarded to the operator that includes any new licence conditions.

3.7.5. Milestone 4: Ready to decommission and remediate a uranium mine and processing facility

Uranium mines and processing facilities that are closed, in active decommissioning or approaching a state of care and maintenance need to have an updated comprehensive environmental management programme to ensure ongoing protection of the environment and the public for both the short and long term during decommissioning, remediation, or care and maintenance [63]. These plans need to follow industry good practice for uranium mine decommissioning and remediation [64].

Mines that are closed or approaching decommissioning need to have an approved and licensed decommissioning and remediation plan that is well structured and ensures that the site is decommissioned and remediated using industry good practices, and that ongoing monitoring is conducted to track the effectiveness of the remediation programme. The decommissioning plan may be separate from the remediation plan and this will be determined by the regulatory framework of the relevant jurisdiction. Previous environmental assessments need to be referred to in terms of decommissioning and remediation outcomes. A comprehensive remediation and end of life plan, complete with monitoring activities, needs to be prepared by the operator and approved by the governing regulatory body. The planned decommissioning tasks that could result in an environmental release (radioactive and non-radioactive pollutants) from the facility and that have an impact on the local environment need to be identified, along with details of the appropriate controls and mitigation measures that are foreseen should such an event occur. The potential pathways that could be associated with these releases need to be described and the potential discharge for each task evaluated.

Once remediation is complete and the subsequent monitoring and surveillance programme have shown that the remediation plans have met the desired results and that the risks to human health and the regional environment have been mitigated, the regulatory body needs to consider removing some or all restrictions placed on the operator for remediation. This may be a graduated approach, where the requirements for surveillance and monitoring are reduced. The last stage would involve the site being fully transferred to institutional control.

Mines that are approaching care and maintenance also need to have a specific, well established environmental management programme that includes treatment of mine water, effluent, tailings and contaminated waste rock water. In addition, mines and processing facilities at this stage need to continue with a well defined environmental monitoring programme commensurate with the risk of the site status and advance towards active reclamation of areas of the operation that will not be used again. Finally, regulatory oversight needs to continue at an appropriate frequency to ensure compliance with the licence conditions prescribed for this stage of a mine's or processing facility's life.

3.8. PROTECTION AND ENHANCEMENT OF CULTURAL, TOURISM, FARMING, PASTORAL AND RELATED INTERESTS

The social and economic interests in an area of uranium exploration, and possible later mining and processing, need to be well understood early in the process. The potential impacts, both positive and negative, become greater if a project proceeds to the mining and processing stage.

While the amount of detail, the effort of the regulatory body and operator, and the interaction with relevant stakeholders are greater for more advanced projects, the aspects to be considered at all stages include the following:

(a) Population density and distribution;
(b) Social infrastructure, including education and health services, formal and informal governance, and availability of a variety of workers and professionals from the workforce;
(c) Physical infrastructure, including transport, water supply, electricity and communication;
(d) Local economic pursuits, such as farming, pastoralism, forestry, manufacturing and other industries, including other mines and tourism;
(e) Conservation and cultural heritage areas and sites.

3.8.1. Milestone 1: Ready to make a commitment to explore for uranium

In some cases, the earliest stages of exploration, such as desktop studies and limited remote sensing, may have little local impact. Once an exploration licence/permit is obtained, however, and on-ground and airborne studies are required, working with local interest groups and communities becomes essential. The housing of exploration crews and their equipment may take up local accommodation resources, or strain recreational, communication, electrical or other facilities, as well as bringing income into an area. The presence of exploration activity in agricultural, pastoral, cultural or tourist areas may also need to be considered.

At this stage, there is a need to interact closely with landholders when the exploration lease extends over pastoral/agricultural land to ensure that the foundations of a good working relationship are set and expectations for impact and remediation are understood before work commences. Damage to fences, crops and pasture should be avoided, and suitable repairs or compensation for crops or pasture that need to be disturbed to allow exploration need to be negotiated and implemented. In some cases, exploration can contribute to the upgrading of existing roads or the creation of new or temporary ones that are available to the community. Heavy use of existing roads or stream crossings may cause damage, requiring additional maintenance work, which needs to be taken into account, and traffic hazards (possibly including collisions with stock or wildlife) may also need to be considered. Although the demand for water is not high during exploration, in arid areas the supply of water for drilling or an exploration camp also needs to be taken into account. Similarly, in more densely settled areas, disturbance to other infrastructure or activities has to be considered.

During the exploration period, many exploration teams may interact with the local community and initiate some local support activities. The extent of interaction and other support will be related to the size and duration of the exploration project. If a deposit is found, advanced exploration and delineation of the potential ore body typically requires significantly more personnel and physical resources than early exploration and would typically be associated with increased support to the local area.

When planned well, some infrastructure improvements implemented for exploration might become assets for the local community. This could include improved roads, water supplies, improved communications or an airstrip.

Exploration projects may consist of several short, seasonal campaigns utilizing only existing local infrastructure, or long term projects requiring a camp with accommodations, vehicle maintenance facilities, sample processing and storage buildings or yards, fuel storage areas and sometimes an airstrip. Even when a long term facility is constructed or rented, the amount of activity can vary, from active drilling campaigns to low key exploration or periods of care and maintenance.

3.8.2. Milestone 2: Ready to commit to developing a uranium mine and processing facility

Should a uranium project proceed to construction and operation (mining and processing), the effect on local infrastructure and society may become more pronounced. In some instances, a new uranium mine may be the main economic activity in an area and needs to coexist with existing land uses, such as farming, forestry, pastoralism, recreation or conservation. The spread of contaminants through streams draining the uranium mining and processing site to potential agricultural plains, where the water is used to irrigate crops, needs to be prevented and considered in the early planning stages of any new mine project.

When landholder agreements are implemented effectively, commitments to local communities and arrangements with the government and regulator are successful. Unavoidable impacts are appropriately compensated (e.g. compensation for lost farmland or houses, provision of alternative roads, diversion of watercourses, replanting of trees or wildlife habitat elsewhere) and an appropriate package of community and societal development is delivered. The

Rössing Foundation is an example of sustained stakeholder involvement focusing on programmes and projects associated with overseeing the Rössing uranium mine's corporate and social responsibilities in Namibia [65]. Additional details on this subject are provided in Appendix I.

The size and budget of a community development package depend on the size of the mining and processing project and local circumstances. Even small mines and processing facilities that cannot support a programme like that of the Rössing uranium mine can have a positive impact on the neighbouring communities. This is in addition to providing some employment and small business opportunities and contributing financially to government income at different levels.

3.8.3. Milestone 3: Ready to operate a uranium mine and processing facility

Much of the discussion in Section 3.8.2 also applies here. However, there may be legacy problems from earlier mining and approaches that require renewed negotiation at long lived mines, such as the Rössing uranium mine in Namibia or the Somaïr and Cominak mines in Niger. In other circumstances there may be legacy sites that have not yet been satisfactorily remediated or were remediated to the standards of some decades ago that are no longer considered suitable. Satisfactory resolution of outstanding legacy sites, or the upgrading of the regulation, operation and preparation for eventual closure of existing (especially long lived) mines, may be an important component in establishing confidence in the capacity of the Member State to enhance its existing capability.

3.8.4. Milestone 4: Ready to decommission and remediate a uranium mine and processing facility

As described in Section 3.8.3, satisfactory resolution of outstanding legacy sites and after care of remediated sites are important at this stage. According to national regulatory requirements, some Member States completely decommission and remediate former uranium mining sites (e.g. France, Germany). Others opt to make selected sites safe and maintain them in a condition such that mining could be recommenced in the future, should circumstances become favourable (e.g. Portugal).

Decommissioning and remediation projects can also lead to improvements in local conditions to compensate partly for the loss of mining employment and income. Some former mining sites retain a heritage value and become education or tourism centres, while others are returned to previous land uses (e.g. farming, pastoralism, forestry) or switched to alternative uses (e.g. recreational, conservational, further industrial use). Clearly understanding what the expected end state obligations or expectations are, before the project proceeds or is expanded, is an important factor. The community and government goals in reaching a long term institutional control status need to be in alignment.

3.9. FUNDING AND FINANCING

The funding and financing requirements for the development of uranium mining and processing, including decommissioning and remediation, are significant[3], and the funding for legislative and regulatory infrastructure development needs to come from government sources. Governments need to understand the commitments involved in developing a uranium mining and processing regulatory programme and the need to develop the broad range of human expertise to manage and regulate uranium mines and processing facilities. This is very important for subsequent efforts to obtain financing for these operations. Gaining the confidence of the financial community requires stable and sustained determination to competently manage the construction, licensing and safe operation of a uranium mine and processing facility.

Initial financing can be pursued in several ways. Total financing and ownership by the government is an option if the nation's economic portfolio provides revenue that can be dedicated to the associated capital and operating costs. This approach might not be feasible for all countries. Export financing is another option for funding, but it provides only for a portion of the overall investment. Local or foreign commercial financing is required to balance the capital cost and to cover the interest accrued during construction. A common approach is to obtain private financing backed by specific government guarantees. Another possibility is to secure private funding by a consortium of partners seeking a return on their investment through revenue generated from the sale of the uranium concentrate produced. Credit worthiness is the first priority for obtaining any project financing. Economic policy, debt management and legal risk sharing mechanisms are all important aspects to be considered when securing financing.

Initially, a Member State needs to consider that the impact of low prices can be highly disruptive to sources of capital and mining operations. Uranium is not

[3] [3]In general, the term 'funding' refers to aspects that are the fiscal responsibility of a government in establishing uranium mining and processing; for example, ensuring that the necessary resources for regulation are provided. The term 'financing' refers to aspects that are the fiscal responsibility of the owner/operator (government or private entity).

priced like other commodities on open exchanges. With only a few buyers and sellers, uranium is commonly marketed through undisclosed long term contracts.

Private sources of capital seek the best return on investment and time. Although the chief executive officer or managing director works for the company, he or she is appointed by the board of directors, who work for the shareholders. Thus, the shareholders, especially those with voting control, have significant influence and the motives of these investors need to be understood. Some investors perceive that the underlying value of the commodity will go up and wish to buy a chance or option on that happening. Caution is needed because such investors do not truly invest in a technical programme. A second type of investor is legitimately interested in the business; some have a penchant for risk in exploration and some seek a return from the mining operations. In general, the latter buy shares of large, dividend paying companies, while the former are primarily interested in capital gains through the appreciation of the company's share price, although they are also partly interested in the underlying option effect. Regardless of motives, investors have a fundamental interest in 'liquidity', which allows them to exit easily without greatly changing the value of the investment.

Partnering with listed companies allows governments to see quarterly financial statements and project descriptions. However, governments partnering with listed companies need to be aware of the legal filing requirements of listed companies. An example is a company listed on the Toronto Stock Exchange with which a government has partnered to explore a new area for uranium deposits. The Member State considers uranium to be a highly sensitive but strategic asset, but its Canadian partner needs to issue press releases announcing material changes and to eventually file geological and engineering reports with the SEDAR [66]. These reports are made available to the public. Some of the company's contracts may even be filed with the relevant stock exchange. In disclosing exploration information, transparency is of prime concern, so that potential investors can make reasoned and informed decisions regarding the nature of a project and the associated risks associated.

Exploration and junior mining companies, whether listed on a stock exchange or private (unlisted), raise capital by selling partial ownership (equity) in the form of shares to investors under the rules of their jurisdiction. Because of the inherent risk in exploration, some regulators apply strict rules to protect investors. For example, in the United States of America, the Securities and Exchange Commission only allows explorers to state a reserve (the part of the resource determined to be profitable to mine). Companies are not allowed to state a resource unless required by another jurisdiction. For example, a Canadian uranium listing that co-lists in the United States of America can state a resource. As a result, the primary listings for companies start on the well known mining exchanges in Canada (TSX, TSX-Venture and recent start-up exchanges such as CDN), Australia (ASX), South Africa and the United Kingdom (AIM) [67–72].

The evaluation of a mining project is conducted in stages because of the inherent risk and capital requirements. At each stage of the exploration — scoping, pre-feasibility and feasibility studies — there is a 'go' or 'no go' decision: whether to proceed to the next stage, put the project on hold or abandon it. CRIRSCO measures the error at the first stage (scoping) to be ±30–40% and that at the second stage (pre-feasibility) as ±20–25%, while the error for a feasibility study is ±10–15% [11]. In a study of mining projects, Bullock [73] found that feasibility studies had errors of between −20% and +27% and cost overruns had a weighted average of 27%.

3.9.1. Milestone 1: Ready to make a commitment to explore for uranium

From a financial perspective, exploration is a high risk activity in the uranium production cycle. It involves many stakeholders, including governments, geological survey organizations, geoscientists, consultants, exploration companies and mining companies. Project funding is different from one actor to another in the development of the project and at different levels. For example, geological survey organizations play an important role in attracting investment in the exploration activity by providing a favourable environment and gathering geoscientific information. Exploration companies focus on the identification of prospective properties or potentially economical deposits, as well as their sale to major companies. Major mining companies focus on the exploration and mining of deposits to generate profits in a sustainable manner.

Before embarking on uranium exploration, the Member State needs to understand that mineral exploration is inherently risky and the probability of finding an economical deposit is very low. For example, Marlatt [74] estimates that for every 1000 exploration projects, approximately one economical uranium deposit will be discovered (a probability of ~1‰). Of these deposits, only one in three will advance to mining through the feasibility stage [75]. In 2016, the estimated 46 year (1970–2016) weighted average base rate for the cost of exploration per kilogram of uranium (indicated) in the ground was about $10 [76].

Figure 3 illustrates the risk and financial expenditure profile for the stages of mining from exploration to decommissioning and remediation. Financial risk is present at all stages of the life cycle of a uranium exploration and production programme, and these risks need to be identified and mitigative measures need to be put in place. As noted, there is a low probability that an area with defined mineralization will advance to an active, positive economic margin mine. Overall, exploration is a high risk and costly endeavour, often with results that do not show a return on investment. A Member State needs to recognize this prior to advancing exploration activities. Once a deposit is located and resources are well defined, the project may advance to an active mine. At this stage, the costs increase as a mine advances into production; however, strong due diligence during the resource evaluation and mine planning stages will greatly reduce the risk of financial loss. Once a mine resource is fully depleted, the Member State needs to recognize that revenue generation will cease and therefore funds need to be in reserve to support and sustain decommissioning and remediation efforts through to completion.

While limited funds from international development agencies may be available, Member States need to provide initial funding in Phase 2. In many countries, regional airborne geophysical and satellite based data have been collected over the past four decades. The government in this situation needs to work to organize these data to build archives that are useful for internal review and perhaps for commercial users, if available, in a secure digital format. On-line cadastre systems allow for governments to provide early stage explorers and prospectors with land ownership and mining tenement information and, in some cases, geological data.

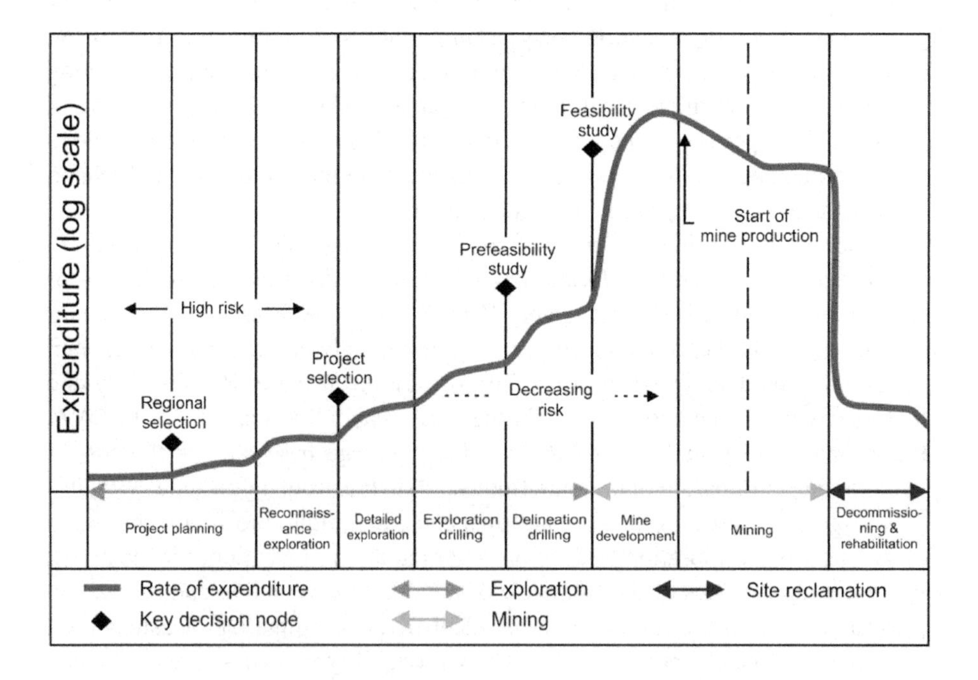

FIG. 3. Risks and financial expenditure for various phases of the life of a mine. Figure adapted from Ref. [77] with permission courtesy of M. Santosh (©2015, China University of Geosciences (Beijing) and Peking University).

3.9.2. Milestone 2: Ready to commit to developing a uranium mine and processing facility

A Member State needs to acquire an understanding of the commitments required for the introduction of uranium mining and processing. It also needs to secure funds early to draft and promote the necessary legislation and the expansion of an existing, or the establishment of a new, regulatory body with the necessary resources to ensure competence. An understanding of the complete life cycle of uranium mining and processing is needed, with specific knowledge of the funding and legislation required for waste management and decommissioning activities. Establishing this regulatory foundation demonstrates the Member State's commitment to advancing uranium mining and processing and will likely be a prerequisite for exploring financing options for these facilities. Construction of a uranium mine and associated processing facility requires significant capital funding. The operator is responsible for securing such funding to construct and commission the mine and processing facility. In some cases, the costs associated with shared or common infrastructure (e.g. roads, electrical or water distribution infrastructure) may be shared between the government and the operator.

The following strategies are needed to support a viable financial plan for uranium mining and processing:

(a) Funding the efforts to create the basic infrastructure necessary to prepare for the introduction of uranium mining and processing;
(b) Developing and maintaining a reasonable level of stakeholder involvement;
(c) Funding for the hiring of expertise to develop the necessary legislative framework;
(d) Funding the expansion or creation of a competent and independent regulatory body and its operation;
(e) Long term financing to ensure the ability to sustain regulatory oversight for decommissioning and remediation of these facilities;
(f) Financing the efforts to support a long term institutional control framework that includes land controls and a document registry of the decommissioned and released sites.

There are various options to source capital after Milestone 1, depending on the levels of risk embedded in both the project and the jurisdiction. This could involve spending of the order of $30 million, depending on the extent of infill drilling required (value from the mid-2010s). Accordingly, some flexibility is required regarding the best path, and generally the financing is best structured in the private sector under industry standard terms. However, some state involvement may be required to facilitate an optimal structure. Governments need to involve experts from the ministry of finance, as well as outside independent consultants, to develop a financial model that can be tuned to consider the stakeholders.

Some large capital expenditure financing in uranium, of the order of $100 million (value from the mid-2010s), has been effected through convertible debentures combining equity and debt, as the value of the loan can be swapped for shares in the company at a future date. However, convertible debentures can be very destructive to a publicly traded company's share price.

Sources of capital with minimum requirements from a history of case studies and varying market conditions need to be considered. Banks and private pools of capital are only appropriate in development projects that are significantly 'de-risked'. The key sources of capital relevant to the uranium industry are summarized below. They show a general 'tolerance' for risk, according to a history of deals. Most commonly, capital financing is implemented through a combination of sources, which is referred to as 'structured financing'. The weighting between sources depends heavily on market conditions. For example, in periods of rising uranium prices the equity market will be more productive.

In Sections 3.9.2.1–3.9.2.3, the following qualifying abbreviations are used:

— E: exploration tolerant;
— R: resources required;
— RR: reserves required;
— OFT: secure off-take[4] desired/required.

3.9.2.1. Sources of capital for uranium exploration

— Capital markets (E):
  • Sophisticated investors;
  • Public companies with capital.
— Producers (E):
  • Major uranium mining companies (e.g. Cameco Corporation, China National Nuclear Corporation, Orano, Uranium One).

[4] [4]An off-take agreement is a contract between a producer and a buyer to purchase or sell portions of the producer's upcoming products. It is normally negotiated prior to the construction of a production facility — such as a mine or a processing plant — to secure a market for its future production. Off-take agreements are used to help the producer acquire financing for future construction, project expansion, or new equipment through the promise of a future income and proof of demand for the product.

3.9.2.2. Sources of capital for uranium resource evaluation

— Private equity (R):
  - Specialist funds will require eventual listing.
— Trading houses (R):
  - Commodity traders, for example, those who buy long term and sell on the spot market.
— State owned enterprises (SOEs)[5] (R):
  - SOEs can be mining or other entities.
— Royalty and streaming companies:
  - Investments made for future royalties and pre-purchase arrangements.

3.9.2.3. Sources of capital for reserve evaluation

— Fuel cycle participants (OFT):
  - Historic interest from enrichment services.
— End users (OFT):
  - Utilities.
— Banks (RR/OFT):
  - Generally high interest will require OFT.
— Merchant banks (R, RR/OFT):
  - Boutique firms specializing in mine project financing.
— Development banks (R, RR):
  - China, South Africa and state owned.

In general, governments need to allow the natural forces of the market to finance mining activities. However, there is a role for government in the following activities:

(a) Coordinating local participation with community leaders early in the project;
(b) Structuring taxes, royalties and SOE participation;
(c) Providing state data and policies during third party due diligence.

[5] [5]'State owned enterprise' refers to a corporation wholly or partly owned by the national or provincial government. Also referred to in some countries as a 'crown corporation' or parastatal.

3.9.2.4. Participation by state owned enterprises

Historically, Member States have had varying levels of participation in commercial mining, both at the national and regional levels. However, over time, governments found that they could weather the typical ‘boom and bust’ economics in the mining industry only if they had a controlling market share through consolidated mine ownership. This happened, for example, in copper, potash and iron ore. Currently, most countries wishing to participate in the mineral sector do so through an SOE. While a minority equity ownership looks appealing politically, it often makes no meaningful contribution to the State unless empowered to off-take at cost and market production. SOE companies can be involved in exploration, but typically do so only if it is self-financed through joint ventures and sales.

An SOE with an active balance sheet may be able to finance an equity position, but Member States need to consider whether they can carry the cost of such capital through to the proceeds of mining. Otherwise, SOE participation is usually through negotiation that is sensitive to the specific factors in the planned mine. Finally, SOEs with a portfolio of assets may be listed on the stock exchange, with the government retaining a significant holding. For legal reasons, the SOE needs to function independent of government.

3.9.3. Milestone 3: Ready to operate a uranium and processing facility

Much of the discussion in Section 3.9.2 also applies here, as the mining company needs to ensure sufficient funding or financing for initial operating costs and provisions for sustaining capital funds. Once the uranium mine and processing facility reach nameplate capacity, the operation needs to, in theory, be fiscally self-sufficient and generate revenue that supports ongoing operational, developmental and sustaining capital costs. At this stage, funding or bonds for decommissioning and reclamation need to be secured from the owner/operator and placed in a national trust fund (or similar mechanism) to ensure that financing is available for these activities. Additional details are provided in the next section.

3.9.4. Milestone 4: Ready to decommission and remediate a uranium mine and processing facility

Industry best practice for social responsibility in the uranium mining industry requires the operator of a uranium mine to continue to operate until the resource is depleted. The operator needs to have protected funds in trust or revenue reserved from production to finance the several years required to effectively decommission and remediate the site. However, several recent events have shown that operators are forced by low metal prices or technical problems to abandon a mine before proper closure. Recommendations to lessen the risk of this situation occurring include the following:

(a) Requiring companies to finance closure bonds as part of the capital requirement up front;
(b) Performing an independent audit of estimated decommissioning and remediation costs, including the option of the operator abandoning the site and the decommissioning and remediation work undertaken by an off-site contractor;
(c) Revisiting the closure commitments and, during periods of improved profit margins, requiring the operator to top up the closure funds;
(d) Inspecting operator filings with overseas regulators (e.g. financial statements, technical reports, news releases);
(e) Analysing mining operations to detect any changes in operations that might indicate financial problems, a faltering operation or changes that could affect the cost of closure;
(f) Requiring companies to finance or secure a bond for any long term institutional control costs once the site is released from licensing.

3.10. SECURITY

The government plays an important role in ensuring nuclear security. The fundamental responsibilities of a government include the following [28, 78, 79]:

(a) Establishing a national security policy and strategy for UOC;
(b) Establishing a legal and regulatory framework for the security of UOC;
(c) Developing a risk based approach to the regulation of the security of UOC;
(d) Establishing, implementing and maintaining a physical protection regime;
(e) Ensuring adequate protection of UOC in use, storage and transport;
(f) Establishing and maintaining a legislative and regulatory framework for physical protection;
(g) Establishing a competent authority responsible for implementing the legislative and regulatory framework.

Security, including physical protection, is intended to prevent malicious acts by internal or external adversaries that might endanger the public, the mine and processing facility employees or the environment. A strong management programme for security, including physical protection, is required for uranium mines and processing facilities. This programme needs to include evidence of a comprehensive review of threats and vulnerabilities and subsequent actions taken to mitigate security risks. Section 8 of IAEA Safety Standard Series No. SSG-60, Management of Radioactive Waste from the Mining and Milling of Ores [80], provides guidance on monitoring and surveillance for facilities that store radioactive materials and waste.

Designing and implementing an effective security programme appropriate for any industrial site containing valuable equipment such as vehicles, high tech equipment, fuel, reagents, explosives and warehouse supplies are key steps. Such a programme includes control or security points to enter the site, and a security gate to exit the site. This ensures that all personnel, goods and vehicles entering or exiting the sites are verified. Any material leaving the site also has to be cleared by the radiation protection department as not contaminated and properly packaged.

3.10.1. Milestone 1: Ready to make a commitment to explore for uranium

Exploration is typically conducted in remote areas and this in itself provides a means of security. Some valuable equipment and supplies remain on-site, so some level of security against theft may be required. Exploration sites need to maintain a level of security to also ensure that radioactive core material does not go missing and does not create a potential contamination/dose risk to the general public or regional biota. During active drilling campaigns, drill core boxes should not be left unattended at the drill site in the field because this would introduce concerns about the security of the drill core, which would otherwise not be guarded and would be vulnerable to damage from domestic animals, human tampering or even destruction by those who may oppose exploration work in their territory. A secure drill core box storage facility (e.g. fenced area, storage room or lockable container) needs to be arranged to ensure sample security. As part of the regulatory process, security procedures need to be in place to track core samples when they are transported between the exploration site and geological or analytical laboratories. Furthermore, the geological and analytical laboratories need a licence to receive, handle and store radioactive substances (e.g. uranium-bearing core samples).

3.10.2. Milestone 2: Ready to commit to developing a uranium mine and processing facility

Once a Member State has made the decision to support and advance uranium mining and processing to the production stage, the requirements for security increase and the following security conditions need to be established:

(a) Legislation providing appropriate authorities for security, including physical protection;
(b) Laws and penalties for criminal activity and malicious acts in or around uranium mines and processing facilities, including theft of nuclear materials or radiologically contaminated materials (e.g. tools, electrical components, vehicle parts);
(c) A site specific security programme that has evaluated security threats and risks and has well defined actions to minimize risk;
(d) A physical protection system that has been tested and received final acceptance from the owner/operator;
(e) A security protocol (including physical protection) for the transport and storage of UOC;
(f) Trained security personnel;
(g) A security culture that recognizes the importance of security requirements for nuclear material.

At this stage, the government and regulatory body need to develop a framework for the protection of UOC [78]. Security considerations include the following:

(a) Defining the state responsibility;
(b) Identifying the competent authority;
(c) Developing the legislative and regulatory framework;
(d) Identifying and assessing threats;
(e) Ensuring that transport security is included;
(f) Defining the responsibilities of the licence holders.

At the initial stages of the development of a mine and processing facility, the owner/operator needs to develop security measures for the protection of UOC [78]. Using a risk based approach, the owner/operator needs to identify the threats and targets and develop mitigation strategies to ensure the security of UOC. In addition, the owner/operator needs to develop a security policy and a corresponding strategy that includes the following considerations of security management:

(a) Security functions;
(b) UOC security culture;
(c) Security plan;
(d) Administrative controls and procedures;
(e) Inventory control procedures for nuclear materials;
(f) Quality assurance;
(g) Information and cyber security.

In addition to the above points, the owner/operator needs to develop management programmes and procedures that include physical protection, inventory control and transport security measures. Consideration also needs to be given to security standards for reporting on uranium production, inventory and exports, as well as security for uranium mines and processing facilities and their workers. A commitment to establishing a strong reporting and security culture needs to be established prior to making a commitment to pursue uranium mining and processing.

3.10.3. Milestone 3: Ready to operate a uranium mine and processing facility

At this stage, the government and regulatory body need to have security policies, legislation and regulatory framework developed and implemented as identified in Section 3.10.2. The regulatory body needs staff fully trained and knowledgeable of the regulatory aspects of UOC security. These aspects need to be implemented and regulated in a practical way, so that they are an integral part of the licence to operate a uranium mine and processing facility. In addition, regulatory inspections and enforcement mechanisms need to be in place that support the regulations prepared for the security of UOC [81].

The owner/operator needs to have all security measures identified in Section 3.10.2 to be fully developed, the relevant procedures detailed, and staff trained to ensure the security of UOC at the start of production. In addition, modern uranium mines and their associated processing facilities are highly automated and contain elaborate process control systems. Computer security of infrastructure and control systems is therefore essential. The owner/operator of these facilities needs to develop programmes, systems and procedures to ensure the computer security of instrumentation and control systems in the mine and processing facility [82, 83]. Security threats may be both external and internal to the uranium mine and processing facility. Prior to starting the operation, the owner/operator needs to conduct a detailed risk assessment of both external and internal security threats and develop mitigation measures [84]. Some examples include theft of uranium in any form found in the mining and processing stages, sabotage of mining and processing equipment or related computer control/instrumentation systems, and sabotage of relevant security systems.

3.10.4. Milestone 4: Ready to decommission and remediate a uranium mine and processing facility

The points defined in Section 3.10.3 are also applicable to this scenario, with the exception that uranium will no longer be mined or processed. However, during remediation, uranium can still be recovered from water treatment or other related activities. During decommissioning and remediation there is reduced staff on-site and possibly new contract employees brought in for decommissioning or demolition work. This further increases the risk of theft of UOC, as well as mine, processing or salvaged equipment and infrastructure, all of which have the potential to be contaminated by radiation. Unapproved removal of such materials needs to be handled as theft, and removal of radioactive contaminated materials from the site would be a regulatory incident and potentially a media related issue. As such, specific regulations and site based operating procedures for site security during decommissioning and remediation are required.

Once the site moves to post-decommissioning and monitoring, and then to institutional control, the level of active security will also diminish. A well designed closure plan will have passive barriers in place (e.g. backfilled pits, closed mine workings, demolished and buried structures) and on-site security personnel or secured gate access should not be required.

3.11. TRANSPORT AND EXPORT ROUTE

UOC produced in uranium processing facilities is considered low specific activity (LSA-1) type material according to section 4 of IAEA Safety Standard Series No. SSR-6, Regulations for the Safe Transport of Radioactive Material [85]. Furthermore, UOC is transported in IP1 packaging. In the uranium production industry, UOC is typically packaged in standard 210 L open head steel drums with a tight fitting lid, which is secured to the drum with a steel locking ring clamped by a locking ring bolt. When full, the steel drums can each weigh between 400 and 500 kg. The UOC is shipped domestically and internationally by road, rail and sea as the UOC is transferred to facilities that refine and convert uranium. The transport of UOC by sea utilizes engineered 20 ft (~6 m) International Organization for Standardization (ISO) sea containers [86]. These containers ensure that the UOC is protected during handling and loading and is further protected from the conditions experienced during sea transportation.

Although UOC is considered LSA-1 type material, it is classified as a dangerous good (Class 7). As such, the international standard for shipment of such material needs to be followed. Each country that transports this material needs to have regulations that ensure compliance with these international shipping standards. Section 5 of the SSR-6 provides guidance on the requirements and transport of UOC [85]. Both regulators and operators who are involved with regulating or transporting UOC need to be trained on these regulations. This also includes training in the transport of dangerous goods. Regulations and transport procedures need to include the requirements and procedures for emergency response in the event of a transport incident involving UOC.

If industry standard packaging and handling procedures are followed, the radiation exposure from packaged UOC is minimal. This is due to the low level of radioactivity of UOC, the short time that handlers are exposed to the shipping container and the stringent packaging requirements. There may be reluctance from transport companies and shipping port personnel to handle Class 7 UOC. Clear and transparent regulations and effective communication and training (including radiation safety and emergency response) are needed to provide confidence and support from these personnel.

Overall, Member States seeking to proceed to uranium production, and subsequently shipping, need to have an understanding of the current industry practices for the packaging of UOC (drums and sea containers), labelling, documentation (i.e. dangerous goods declaration, radioactive monitoring record, export licence, import licence, transport documents, safety certificate and required transit licences), container security seals and transport logistics. Regulations need to be in place for the transport of UOC and comprehensive training programmes are needed to ensure that regulators, operators and shippers/handlers are appropriately trained.

With regard to the transport route, uranium mines are sited where uranium deposits are located, and the processing facilities are usually nearby. They may be in a remote location and at significant distances from standard transport routes. As such, transport routes and licensing conditions need to be considered prior to the development of a uranium mine and processing facility, including accommodations for road or rail transport from the processing facility, as well as a port for shipment of UOC by sea.

3.11.1. Milestone 1: Ready to make a commitment to explore for uranium

As stated above, exploration for uranium typically occurs in proximity to where the uranium is geologically located. Exploration sites are often located in remote sites where there may be inadequate roads or other transport infrastructure, so exploration projects are typically self-supporting. Therefore, well developed transport routes are not required for early stage exploration projects. Exploration projects can rely on air transport of materials and equipment required to explore for uranium. In addition, off-road vehicles may be utilized to access remote exploration areas. Planning for an exploration project needs to include the transport of materials and equipment in and out of the exploration area. Tree clearing, stream crossings or ground disturbance permits may be part of the licence to explore, and an exploration company may be required to define areas impacted during the transport of equipment and the site itself prior to issuance of an exploration licence. Better developed transport routes, including tree clearing and installation of dirt roads and bridges, may be required for advanced exploration projects, requiring enhanced environmental reviews and permits.

3.11.2. Milestone 2: Ready to commit to developing a uranium mine and processing facility

An increased focus on transport and export routes is required at this stage. If land preparation to develop either an underground or open pit mine is needed, then heavy earth moving equipment will be required. Suitable ground or sea freight transport is required to move such large equipment. In addition, the development of an underground mine and processing facility will require the transport of large processing equipment and infrastructure (including cranes to facilitate the unloading and placement of equipment and infrastructure). Therefore, well defined transportation routes will have to be in place prior to advancing to the development and construction of a uranium mine or processing facility. Effective and sustained transport and export routes will also be required to move materials and bulk commodities to the site during operation. This includes importing bulk commodities and materials from outside the country, if required. In addition, a mode of exporting the final UOC from the site that is in keeping with national and international standards will have to be considered in advance of the design and development of a uranium mine or processing facility. Finally, the operator and Member State need to ensure that transport carriers (i.e. trucking, rail, shipyards/ports and shipping companies) are licensed and certified to handle and transport UOC.

3.11.3. Milestone 3: Ready to operate a uranium mine and processing facility

At this point, the required transport routes for materials, supplies and bulk reagents to the site and UOC from the site need to be well defined and construction needs to be completed. Licensing and training requirements for the shipment and handling of UOC need to be well defined and in place.

A Member State that has a long history of uranium processing and wishes to enhance existing capacity and capability has to assess current transport and export infrastructure and determine whether it can support, from a capacity and logistics perspective, the current and future uranium production requirements. This includes roadways, shipping ports, transport hubs, off-site warehouse or storage areas, and rail facilities. Uranium mine and processing facility operators and Member States need to work with local transport companies to ensure support for the transport of materials and bulk commodities on the basis of current and future needs.

3.11.4. Milestone 4: Ready to decommission and remediate a uranium mine and processing facility

The transport and export infrastructure that was in place during the operation of the uranium mine or processing facility needs to be able to support decommissioning, remediation, and care and maintenance activities. This type of infrastructure must be considered during the planning for decommissioning, remediation, or care and maintenance. During the final stages of decommissioning, some site access roads may also be decommissioned, if that forms part of the site closure plan. Access roads, air strips, stream crossings and even dedicated seaports to that project need to be reassessed to determine whether they should be kept or removed. The needs of the community and the government will be an important part of those final decisions.

3.12. HUMAN RESOURCES DEVELOPMENT

The knowledge and skills necessary to locate a uranium deposit and then design, construct, license, operate, maintain, decommission and remediate a uranium mine and processing facility include all aspects of scientific, engineering, administrative, financial and management disciplines. While much of the knowledge and many of the skills required are the same for any exploration project, mine or processing facility, there are specific considerations for a uranium mine and processing facility [87]. Appreciation is needed of the increased attention to detail necessary to ensure operational safety, security and radiation protection, in particular as the uranium becomes more concentrated as processing moves upstream and the radioactive waste accumulates. Specific expertise in design, operation and maintenance are required to ensure effective radiation protection for mine and processing facility workers.

Human resources development is a complex task and may vary widely, depending on the national decision at each stage of exploration, construction and operation as to whether to fulfil needs through indigenous development or by procuring capabilities from outside the country. Even if procuring outside human resources is the preferred approach initially, developing domestic capabilities may be considered for the long term. The development of such domestic capabilities may require a significant focus on education and training. This may include education and training programmes that are supported through governmental or academic institutions.

Industry good practices for workforce development and training at a uranium mine and processing facility need to be followed to ensure safe and efficient production [88]. This includes development of the management and worker structure of a uranium mine and processing facility (i.e. mining, processing, maintenance, engineering, environment, health and safety, human resources, security and administration). Pre-employment and site based training strategies need to be developed and implemented prior to commissioning and operating a uranium mine and processing facility. Site based training includes new employee orientation, environmental and safety training, radiation protection, first aid, emergency response, mine rescue and firefighting, mine operations training, processing facility training, and trades and technical training.

Accordingly, the knowledge, skills and training required to effectively regulate uranium mines and processing facilities need to be developed in a Member State prior to construction, commissioning and operation. Once the standards and regulations for uranium mines and processing facilities have been put into place, the national regulatory body needs to be managed and staffed by competent technical personnel [42]. A wide variety of activities are involved in regulatory oversight, from exploration to decommissioning, so the regulatory staff need to be competent in a number of technical disciplines. This includes setting of standards, administrative procedures for licence applications and reviews, oversight of operations, inspections of facilities, enforcement, requirements for confidentiality, record keeping and public/stakeholder information. Therefore, the regulatory body needs to have staff with diverse educational backgrounds and general to expert knowledge in a number of disciplines, including health physics, radiation protection, conventional safety, mining, hydrometallurgical uranium processing, environmental management, geochemistry, hydrogeology, inspection protocols, legal aspects and record keeping. In addition, as part of the on-boarding process, the staff of the regulatory body have to undergo specialized training that includes some form of certification (because of potential legal challenges) to effectively administer and enforce the regulatory programme. This requires significant organizational efforts, time and funding, as well as a well qualified regulatory body. Soliciting expertise from the IAEA to assist in the development of this regulatory training needs to be considered.

3.12.1. Milestone 1: Ready to make a commitment to explore for uranium

In general, exploration for uranium can be conducted by a dedicated exploration company, by a uranium mining company that has a dedicated exploration department or by a government run geological survey. If a government geological survey employs its own dedicated exploration team and a decision has been made to explore for uranium, competent and trained personnel needs to be available to conduct all phases of the exploration activities, including highly technical resource modelling and estimation.

If a private exploration company locates an economical and recoverable uranium deposit, it will typically try to sell the deposit to an existing uranium mining company, a startup mining company wishing to begin mining and processing of uranium or a Member State seeking to extract the resource for domestic nuclear fuel sources, and might try to manage and conduct the operation through a division of the government. Under these circumstances, the Member State does not need to focus on the development of human resources, as exploration is typically a project based activity and should not be viewed as an activity that sustains employment. In terms of regulatory oversight, the Member State's existing mineral exploration review and approval programme is sufficient, with only some minor enhancements.

A third option may be where a government run geological survey completes exploration activities for uranium. If a government employs its own dedicated geological survey exploration team and a decision has been made to explore for uranium, then it needs to ensure that it has competent and trained personnel to conduct all phases of exploration activities, including resource modelling and estimation. Finally, the human resources required for the development of regulations and for regulatory oversight of uranium exploration activities need to be considered.

3.12.2. Milestone 2: Ready to commit to developing a uranium mine and processing facility

If an economical and recoverable uranium resource is identified in the Member State, a uranium mining company or a government may seek to develop that mine. At that point, there needs to be a review of local human resources capabilities and an evaluation of whether suitable domestic resources are available for construction, commissioning and, ultimately, operation. If the Member State does not have domestic human resources capacity capable of designing, constructing and operating a uranium mine or processing facility, then the mining company will have to recruit employees from other relevant national industries or recruit employees from outside the Member State. In consideration of the life of mine, the Member State may work with the mining company early in the project to develop a domestic workforce through training and development programmes. This is one of the key aspects of sustainable development that a Member State may consider as a foundation.

At this stage, the physical setting and logistics associated with the workforce at the mine and processing facility needs to be considered. This includes determining whether the employees will be housed in a camp facility at the mine or transported to a site at some predetermined frequency. It is important that human resources issues be considered at the same time that the mine and processing facility are being designed [88]. This will provide sufficient time to ensure that a well qualified workforce is in place when the facility advances to the commissioning phase and, ultimately, full operation.

The Member State also needs well qualified regulatory personnel who will develop the regulations, codes and standards by which the uranium mines and processing facilities will be licensed and regulated. The regulatory team needs to have some basic working level knowledge of uranium mining and processing and of the different stages involved, including siting and construction, operation and decommissioning. This competence is required to ensure effective regulatory oversight.

Other aspects of human resources requirements that need to be considered as a Member State proceeds to uranium mining and processing include the following:

(a) Political and social expertise for public communication and consultation.
(b) Technical and regulatory expertise to develop and implement regulations, codes and standards for uranium mines and processing facilities. This includes the licence to construct, the licence to operate, radiation protection, conventional safety, emergency planning, oversight of management programmes, environmental management, waste management and decommissioning.
(c) Expertise to conduct training programmes for operations and maintenance personnel as the uranium mine and processing facility transitions from construction to commissioning and to operation. A training needs analysis is required that summarizes the training requirements for each position in the mine and processing facility. A decision is needed as to whether ancillary services will be conducted at the mine and processing facility or outside services will provide support (e.g. payroll, accounting, procurement, engineering, non-routine maintenance functions).

(d) Strategy and plans to develop and train the regulatory body required for construction and operational oversight.

3.12.3. Milestone 3: Ready to operate a uranium mine and processing facility

At this stage, a Member State has to recruit and train regulatory staff to regulate the uranium mining industry. In addition, the regulator needs to ensure that an effective training programme has been developed by the operator and that mine and processing facility employees are trained to ensure safe commissioning and operation. Evidence of an effective training programme and of trained and certified mining and operational personnel needs to be demonstrated as part of the issuance of the licence to operate. As part of the improvement cycle, an operator needs to continuously evaluate the requirements of the facility and refine its human resources requirements and training plans based on both internal and external factors, such as audits.

Finally, a Member State that wishes to enhance its existing capacity and capability needs to complete a gap analysis to determine what additional human resources, if any, are required to increase capacity and capability. If the production capacity at an existing operation is reached through the installation of larger equipment, then an increase in employees might not be required. If a current mine is being expanded or a new mine is being developed, then additional human resources may be required. Under these circumstances, a Member State needs to look at opportunities to transfer some of its experienced employees to the new facility to assist with training, commissioning, startup and operation.

3.12.4. Milestone 4: Ready to decommission and remediate a uranium mine and processing facility

A Member State with sites that are either closed, in decommissioning or remediation, or moving into care and maintenance with the intention to restart at some point, will require a different level of expertise than an operating site. A significant proportion of the skills that are developed during operation are transferrable to this stage, and workers with this operational experience can assist in decommissioning or reclamation activities. A needs analysis ought to be completed to assess the quantity of workers that are required and the skills that they require to support this stage. Next, a gap analysis needs to be completed on the workers selected to support this stage, and a training programme has to be developed to ensure that they have the required competencies. Some decommissioning activities, such as demolition, require specialist skills that may be brought to the site by a contractor. Typically, a smaller workforce may be required for decommissioning and reclamation than was required during the operational stage. A mine or processing facility that is going into care and maintenance will probably still require some operations, maintenance, technical, administrative and managerial personnel to maintain it in a state where it can meet the regulatory requirements for operation at this stage. In addition, active remediation of areas of the operation that will no longer be used could be mandated by the regulatory body, which may require specialized personnel or consultants to complete design and implementation.

3.13. SITE AND SUPPORTING FACILITIES (INFRASTRUCTURE)

Site selection and the evaluation of an exploration area, uranium mine or processing facility are constrained by the location of the uranium ore deposit. For this reason, supporting facilities typically need to be developed to support the mine or processing facility. This can add complexity from a financial perspective (e.g. roadways, transportation links) and a geopolitical viewpoint (e.g. source of water for the mine and processing facility, potable water, disturbance of pastoral land). A Member State needs to consider all aspects of supporting infrastructure prior to providing a licence to construct a uranium mine and processing facility. Because mining is a temporary use of the land, the government needs to be cautious about locating a new permanent town site near the facility, as single industry towns are vulnerable. Some aspects are described in more detail in the following sections.

3.13.1. Milestone 1: Ready to make a commitment to explore for uranium

An exploration project for uranium is generally self-sufficient with regard to its need for infrastructure. Most such projects operate in remote locations and thus need to be self-sufficient in terms of power, fuel, water, communication, food and shelter. In addition, exploration areas are usually accessed using off-road vehicles or helicopters.

3.13.2. Milestone 2: Ready to commit to developing a uranium mine and processing facility

Preliminary investigations regarding the required infrastructure and supporting facilities need to be made during the design of the mine and processing facility and should continue during detailed engineering. Some of the fundamental considerations include electrical requirements and the availability of electricity at the mine site during construction and operation. An assessment of electricity demand for the mine and processing facility needs to be completed, including aspects such as whether electricity will be generated on-site by generators as an initial option and/or later as backup power. If the state electricity grid is used to service the mine and processing facility, upgrades to the power station, substations and transmission lines may be required if they are determined to be inadequate. In addition, road or rail infrastructure is important to transport construction and operating materials (including bulk reagents and fuels) to the mine site. Water supply from both an industrial and a potable perspective is also an important consideration. Fire protection systems are required for protection of infrastructure, human health and the local biota. Permanent maintenance, materials acquisition and warehouse facilities need to be considered to support construction activities, as removal of such facilities and subsequent construction of permanent facilities can be costly and generate waste.

Access to health care for sick or injured workers also needs to be considered during initial mine design. In certain jurisdictions, access to adequate health services is part of a mine's operating licence. The proximity of the mine to local communities will dictate whether a mining camp facility or dedicated mining town complete with services will be erected. Mining camps or subsidized mining towns can add considerable complexity and cost to a mining operation and may be only temporary. The operator and the government need to consider these options early in the mine development as part of the life cycle costs for the operation.

Another issue is the need for an air strip nearby. Depending on how isolated the site is, the quality of the roads and how far the workers have to travel to reach the site, an air strip or airport needs to be considered. As this may also benefit the local communities, an airport could become viable and could remain even after the end of the mine's life.

3.13.3. Milestone 3: Ready to operate a uranium mine and processing facility

By this stage, all site and supporting infrastructure needs to be in place and testing completed to ensure that the infrastructure is functional and meets peak demands. A uranium mining or processing operation wishing to increase its current capacity needs to review the capacity of its current infrastructure. One immediate aspect to review is the supply of electricity, in comparison with the forecasted peak demand should there be a change in the operating infrastructure that places additional demand on the local grid. Additional generator capacity may be required, or the operator may have to discuss forecasted electrical demands with the local utility to ensure that future demands are met. Another important aspect is water demand from a processing perspective. An operator has to determine whether an increase in capacity is equivalent to an increase in freshwater requirements, as well as wastewater volumes or waste management volumes, such as tailings. In that case, modifications to the regulatory permit may be required prior to utilizing additional fresh water in the operation or expansion of the waste management system. If the increased capacity results in additional mine or processing facility employees, then the operator needs to assess the capacity of the mine camp or local community. Additional housing may be required to house the extra employees.

Overall, the operator would have to complete a condition based analysis of the infrastructure in place and determine whether it requires upgrading or replacement prior to resuming operation or increasing production. In addition, for operations that were previously in care and maintenance, the regulatory or socioeconomic standards may have changed, depending on how long the operation was off-line. The operator then needs to complete a gap analysis to determine any shortcomings in infrastructure prior to restarting the operation.

3.13.4. Milestone 4: Ready to decommission and remediate a uranium mine and processing facility

A Member State that has uranium mines or processing facilities that are closed, at the end of life or in care and maintenance requires the same level of infrastructure as that of an operating mine or processing facility. Activities involving active remediation and water treatment require power, mobile equipment (e.g. fuel, spare parts) and trained personnel. Therefore, good road infrastructure is required to continue to transport parts, reagents and food to the site. In addition, the electricity grid and associated infrastructure is needed for water treatment facilities under these conditions. Finally, a reduced version of the camp facility (if present) is required to support the staff remaining during closure or care and maintenance.

In general, it is unlikely that any supporting infrastructures will remain in place during the post-decommissioning monitoring period or the long term institutional control phase. This changes if the government or the community identifiess key features that they wish to maintain or utilize (e.g. airstrip, communications tower).

3.14. CONTINGENCY PLANNING

Contingency planning is important for exploration projects and operating uranium mines and processing facilities. Risks need to be evaluated from all perspectives, and monitoring programmes and mitigation measures need to be developed and implemented to ensure the protection of the workers, the general public and the environment in the case of unusual natural events (e.g. an infrequent high rainfall event, earthquakes, far ranging pandemic type health issues), sociopolitical conditions (e.g. security requirements for civil unrest) or catastrophic failure of mine or processing facility infrastructure.

3.14.1. Milestone 1: Ready to make a commitment to explore for uranium

Uranium exploration activities typically take place in remote locations and possibly in challenging climatic conditions, such as extreme heat or cold. An exploration project needs to be able to operate independently in remote locations, and therefore those involved need to have geographical, climatic and political knowledge of the location. They therefore need to be prepared to deal with changing conditions and to have contingency plans should conditions change, prompting an evacuation of the exploration area. In addition, contingency plans need to be in place should there be a significant health or safety incident at the exploration site that requires emergency medical attention or evacuation due to natural disasters (e.g. fire, flooding).

3.14.2. Milestone 2: Ready to commit to developing a uranium mine and processing facility

A mine that is under development needs to complete a comprehensive risk assessment and have detailed, well documented contingency plans in place to deal with significant process upsets (e.g. mine flooding), extreme weather conditions, loss of electrical power, forest fire, outbreak of illnesses at the site (e.g. pandemic), civil unrest or fire at the mine site. The development of protocols for an emergency command centre need to be considered for emergencies, with special attention to communication and updates to the regulatory bodies and other concerned stakeholders. The site needs to have a detailed contingency plan that clearly describes the mitigation measures in place to deal with emergency situations. A detailed description of the safe shutdown of the mine and processing facility needs to be included in the mitigation measures. The regulatory body needs to be informed about the operation's risk identification and mitigation strategy as part of contingency planning and the operator needs to show evidence of this as part of the licencing requirement to construct and operate a uranium mine.

3.14.3. Milestone 3: Ready to operate a uranium mine and processing facility

The measures for contingency planning as described in Section 3.14.2 need to be well documented prior to final commissioning and operation. A list of contingency issues must be identified in the environmental assessment process and initial siting licences. Once the mine and processing facility are active, the facility operator needs to review its contingency plans on a frequent basis to ensure that the plan remains current and supports the needs of the operation. Changing conditions may introduce new operational or security risks to an operation and these risks need to be evaluated and mitigation measures implemented as part of the contingency planning process. Examples include installation of new mining or processing equipment or a power failure affecting worker health and safety and local communities (e.g. failure of tailings dam dewatering pumps). Also requiring testing, at regular intervals, is the emergency response system and its emergency backup power generators. This needs to include evaluation of site management, the site emergency response team, local authorities and communication processes with internal and external stakeholders.

3.14.4. Milestone 4: Ready to decommission and remediate a uranium mine and processing facility

The principles described in Section 3.14.3 for an active uranium mine or processing facility also apply to an operation that is being actively decommissioned, in care and maintenance or under active remediation. Under these scenarios, there are typically fewer site personnel present compared with an active operation, and therefore security and control of the site need to be considered, in particular for emergency situations. A mine operation in care and maintenance or active remediation needs to have an updated contingency plan with mitigation measures that are ready to be acted upon in case of an emergency. In addition, with a potentially reduced workforce, site personnel may be required to play several roles during an emergency situation. The contingency plan needs to provide a detailed role description, as well as the training required for each position.

3.15. WASTE (INCLUDING TAILINGS) MANAGEMENT AND MINIMIZATION

Waste from uranium mining and processing presents a potentially significant risk and long term environmental liability to the mine operation if it is not managed properly. A robust, well financed and enforceable operations plan for the management (including segregation, storage, treatment and disposal) of waste generated by uranium exploration, mining and processing needs to be a mandatory requirement for the issuance of a uranium licence by a regulatory body for uranium exploration, development, commissioning, operation, decommissioning and remediation. The operator needs to submit for approval, where required, comprehensive and detailed guidelines for the management of radioactive waste across a project life cycle. In addition, the operator needs to account for and report on the quantities of all types of radioactive and non-radioactive waste produced and provide information on their management.

From a due diligence perspective, a comprehensive extraction approach needs to be developed, where an operator seeks to maximize the extraction of all resources that are economically beneficial, thereby maximizing the resource and minimizing the waste produced. This concept needs to be aligned with an appropriate waste management policy developed by the operator that includes minimizing the site footprint; disturbing the ground only once during mining and extraction; optimizing returns from the valuable materials in an ore body; integrating primary and secondary resource management for resource conservation and waste prevention; segregating the waste materials and reusing clean mine rock materials for construction purposes; fostering other types of reuse; and recycling and new product development (i.e. from recycling tailings or residues) in line with the waste minimization hierarchy, where all waste needs to be properly managed to reduce long term negative environmental liabilities.

The government needs to have a clear understanding of the requirements for the management of waste generated from uranium exploration, mining and processing activities. If this regulatory framework has not been established in the Member State, then the government needs to refer to regulatory requirements for the management of waste in countries that have active uranium exploration, mining or processing operations and use industry accepted best practices for the management and regulatory oversight of the various types of waste from these activities.

3.15.1. Milestone 1: Ready to make a commitment to explore for uranium

A Member State that has current exploration activities or intends to explore for uranium needs to develop regulatory based waste management standards based on international good practices. An exploration company that is actively drilling to identify or delineate uranium deposits needs to restore the exploration site to its original state. This includes management of non-radioactive waste and infrastructure. It also includes any radioactive waste associated with this type of activity. This could include radioactive drilling muds, processing water contaminated with radioactive elements, or a radioactive drill core. The regulatory body needs to develop guidelines on managing and disposing these types of waste and provide oversight to ensure that exploration companies remain compliant.

3.15.2. Milestone 2: Ready to commit to developing a uranium mine and processing facility

Management of radioactive waste at a uranium mine or processing facility is a complex activity and requires a high level of technical competence of the operations staff and regulatory personnel. Several types of radioactive waste can be generated from uranium mines and processing facilities, and those with the largest volume are typically mine rock, tailings and contaminated waters. These types of waste present the greatest environmental challenge to a uranium mining and processing facility. Accordingly, contaminated waste rock management, wastewater management and tailings facilities need to be engineered using industry good practice to ensure long term geotechnical and geochemical stability (e.g. for 10 000 years). Furthermore, these facilities need to be managed and regulated carefully from a civil, geotechnical, geochemical, environmental and hydrogeological perspective. External technical support for both the operator and regulator may be required to ensure that these significant environmental liabilities are well designed, monitored and managed [58].

During the design of the mine and processing facility, attention is needed with regard to handling and containment of radioactive process slurries and liquid process solutions. Engineering designs need to feature proper containment, namely both primary (e.g. tanks, piping) and secondary (e.g. berms, bunds, sumps) containment. In addition, control of radioactive dust from areas such as mine rock piles, ore stockpiles, tailings facilities, crushing and grinding plants, and uranium dryers/calciners need to be considered in the design. Effective monitoring and maintenance programmes for this infrastructure need to be developed prior to commissioning and operation.

Other types of radioactive waste may also be generated from uranium mining and processing, including radioactive mine and processing facility infrastructure (e.g. process piping, tanks, pumps, electrical components, building infrastructure, wooden pallets). Industry good practice is to remove the radioactive contamination (where practical) so the materials can be recycled or disposed of in a domestic landfill facility. The regulatory body needs to develop standards and criteria for the management (including reuse or recycling, where applicable) and disposal of these types of waste [58].

Finally, radioactive slime, sludge and residues will also be produced during mining and processing of uranium. Regulations are needed to manage these types of waste by working to incorporate them back into the process, storing them in the mine workings, or geochemically stabilizing them and storing them with the final tailings. Alternatively, if appropriate and practicable, they could be stored in an approved low level waste repository.

Overall, the management of radioactive waste is a complex process in uranium mines and processing facilities and requires considerable attention during the environmental assessment process. The operator needs to demonstrate to the regulatory body that qualified staff and, if required, technical consultants work collaboratively to ensure that radioactive waste materials are properly managed and disposed of. Radioactive waste originating from a uranium mine and processing facility present the greatest long term environmental and social liability associated with these types of activities. Finally, the regulatory body needs to have the requisite technical expertise to ensure that the risks associated with waste streams generated on-site have been correctly identified, strategies for radioactive waste management are scientifically competent, monitoring programmes are effective and risks are being mitigated to ensure that people and the environment remain protected.

3.15.3. Milestone 3: Ready to operate a uranium mine and processing facility

At this stage, there needs to be a technically competent regulatory body in place that has a well developed set of radioactive and non-radioactive waste management standards and regulations based on industry good practice. The facility operator needs to have a comprehensive waste management programme, plans and associated infrastructure in place prior to the final commissioning and operation of the mine and processing facility. From a regulatory perspective, evidence of these needs to be clearly demonstrated before a licence to operate is issued. The regulatory body also needs to have a compliance programme in place commensurate with the risks identified from the operation in terms of waste management to support this approval.

A Member State wishing to enhance existing capacity and capability with uranium mining and processing needs to evaluate current practices and regulatory standards. These have to be comparable with international industry good practices for the management of radioactive waste. Operators seeking to increase capacity and capability need to look for opportunities to mitigate the risk of long term environmental and social liability associated with the management of radioactive waste. This again needs to be based on international industry good practices. The operator needs to implement these good practices if they are reasonably and practically achievable.

3.15.4. Milestone 4: Ready to decommission and remediate a uranium mine and processing facility

Safe and efficient management of radioactive waste generated from uranium mining and processing is critical at this stage. If it is ineffective, the result can be significant environmental and social liabilities that may be very expensive and require several generations to ameliorate [34]. Decommissioning and remediation plans that are technically sound and executable need to be developed by the operator and approved by the regulatory body before decommissioning and remediation can begin. Opportunities to decommission and or remediate inactive mine or waste areas during the operating life of the facility need to be encouraged, as decommissioning and remediation liabilities and costs can be minimized sooner. Reviewing the recommendations from the environmental assessment for the application to develop the operation needs to be performed by the operator and the regulatory body. If the project was started with the end in mind, the final planning objectives will be easier.

Decommissioning of mine rock piles, mine workings (open pits or underground mines), in situ recovery well fields and tailings facilities is a complex activity that requires an expert level of knowledge and coordination from a civil, geotechnical, geochemical and hydrogeological perspective. The operator needs to have the necessary technical expertise in the organization or to hire competent consultants to facilitate these activities. The operator needs to classify all radioactive waste and develop waste management options. Similarly, the regulatory body needs to have the technical competence to critically review the decommissioning and remediation plans proposed by the operator, including radioactive waste management strategies. The regulatory body is also accountable to the government for effective oversight of the decommissioning and remediation of the uranium mine or processing facility, as it will ultimately approve the decommissioning and remediation plans.

Mines or processing facilities considering care and maintenance with the intention of restarting production later need to effectively manage their radioactive waste and waste water. This may include active dewatering and treatment of contaminated water originating from contaminated waste rock piles, the processing plant, mine water and tailings facilities to ensure that the impact on the environment from these facilities is mitigated. In addition, operators should be encouraged to either clean up or remediate all sources of radioactive waste during a care and maintenance period to ensure that the environmental liability remains ALARA.

3.16. INDUSTRIAL INVOLVEMENT, INCLUDING PROCUREMENT

Many commodities (including bulk reagents and fuels), operating and infrastructure components, and services are required to construct and support the operation of a uranium mine or processing facility. Mining equipment, both fixed and mobile, processing piping and equipment, spare parts, consumable supplies, fuel, safety equipment, specialist contractors and instrumentation components are some examples of industrial involvement supporting a uranium mine or processing facility. Several of these examples are common to the mining and processing industry from a general perspective, and if a Member State has a history of mining and processing of other metals this type of support may already be present. Industrial involvement can be a source of employment and economic growth in the country and region where the mine or processing facility is located.

3.16.1. Milestone 1: Ready to make a commitment to explore for uranium

Exploration organizations are generally self-sufficient but may require some industrial involvement during all stages of the exploration activity. This may include procuring aircraft for airborne surveys, support for exploration camp operation (local contractor), rental vehicles or supplies for drilling programmes (i.e. drilling equipment, oils or drill mud).

3.16.2. Milestone 2: Ready to commit to developing a uranium mine and processing facility

At this stage, the Member State and facility operator need to jointly assess the local and national industrial support capabilities for the uranium mine and processing facility. The Member State needs to encourage the mine or processing facility developer and operator to procure industrial supplies, services and commodities from local or national suppliers as a means of retaining both the indirect and direct economic benefits of uranium mining and processing in the Member State. This may include timely delivery of high quality construction materials required to construct a uranium mine or processing facility and to keep the project on schedule. The owner/operator needs to determine the level of industrial support required during operation and work with local and national suppliers to determine whether they have the capability to consistently supply commodities, components and services to the uranium mine or processing facility. Unique activities that arise may include in situ recovery well field construction, shaft sinking for underground mines, hoisting operations and pressure leaching vessels for processing. The procurement department of the owner/operator has to meet with individual local and national suppliers and evaluate their capacity to support the construction and operation of a uranium mine or processing facility. Once a decision is made to construct a uranium mine and processing facility in the country, the Member State needs to encourage local and national suppliers to be proactive and develop strategies to support these operations, including co-partner relationships with external suppliers. This will ultimately support employment and provide economic benefits to the local and national economies.

3.16.3. Milestone 3: Ready to operate a uranium mine and processing facility

At this stage, procurement supply strategies and contracts need to be developed to support the operation at its nameplate capacity. This is an important aspect to ensure safe, reliable production including robust environmental management of the site (e.g. secure supply of water treatment reagents). Logistical challenges and increased demand/supply opportunities need to be considered, as an operation may require more supplies than what was forecast during original design, as well as ready access to replacement parts over the life of the operation. With some of these contracts comes the need for not only the materials but also the expertise to construct, commission and operate them. Once operational, the hand-off to the operator from the contractor would fulfil the procurement and supply strategy.

3.16.4. Milestone 4: Ready to decommission and remediate a uranium mine and processing facility

The strategy for local and national industrial suppliers described in Sections 3.16.2 and 3.16.3 also applies to this scenario. Fewer mining and processing components, supplies and services will be required at this stage in the life cycle of a mine or processing facility, although some unique decommissioning and demolition supplies may still be required. The owner/operator of the mine or processing facility needs to consult with the industrial supplier to determine what materials and supplies will be required for decommissioning, remediation, or care and maintenance. In addition, fewer warehousing personnel may be present at the mine or processing facility under these conditions, so industrial suppliers may be required to provide increased inventory management support and 'just in time delivery' based on customer requirements.

4. CONCLUSIONS

National infrastructure that is well established to support the development and life cycle of a uranium mining and processing operation will facilitate safe and sustainable production and encourage strong stakeholder support. It is important to stress that the construction, operation and decommissioning of a uranium mine and processing facility are complex and require a well defined and implemented national policy and regulatory framework, as well as safety, environmental and technical infrastructures. The development, operation and decommissioning of a uranium mine and processing facility may span several decades, and a Member State needs to ensure that appropriate financing and governance mechanisms are developed and sustained through the life cycle of such facilities. Addressing the activities identified across the five phases of the uranium production cycle to achieve the four milestones identified in this publication in a systematic manner will ensure that uranium mines and processing facilities can be regulated and operated in a safe, efficient and environmentally sound manner.

Appendix I

CASE STUDY OF NAMIBIA

I.1. BACKGROUND OF THE URANIUM INDUSTRY IN NAMIBIA

I.1.1. Overview of operating mines and mines in development

Uranium minerals were first discovered in Namibia in 1928, in the vicinity of what would later become the Rössing mine [89–99]. However, it was in the 1960s that Rio Tinto acquired exploration rights in the area and subsequently discovered a number of low grade uranium ore bodies on the north side of the rocky Khan River. Detailed beneficiation test work was carried out, with the results leading to the opening of the Rössing mine in 1976. With more than 40 years of operation, this is currently the longest operating uranium open pit mine in the world. A notable increase in the demand for nuclear fuel for power generation in the 1960s and 1970s, as well as the successful development of the Rössing mine, led to an intensification of uranium exploration in central Namibia by a number of different companies. However, the slow decline in uranium prices prevented the uranium deposits that were identified during the course of this exploration from being developed into operational mines for several decades. Increasing uranium prices at the beginning of the new millennium eventually resulted in the establishment of the Langer Heinrich mine in 2007, when uranium prices were at an all-time high, which once again led to an exploration rush in the western Erongo region. The Langer Heinrich mine produced 1526 t U) in 2017. In the same year, the Rössing mine produced 1790 t U. High resolution airborne geophysical data from the Geological Survey of Namibia were key in exploration efforts to detect the world class Husab uranium deposit, which now supports the Husab mine, set to become the world's third largest uranium producer. Following the production of its first drum of yellow cake on 20 December 2016, the Husab mine produced 1140 t U in 2017. The mine reached its annual nameplate capacity of 5680 t U in 2020. Figure 4 illustrates the Erongo region and the developed uranium mines in Namibia. A summary of these mines is provided in Table 2 and a summary of advanced uranium exploration projects is given in Table 3.

The projects of Zhonghe Resources, Valencia Uranium, Bannerman Resources, Reptile Mineral Resources and Exploration, and Marenica Energy are at advanced stages of exploration and optimization test work for uranium extraction. The full development of these projects awaits an increase in the price

TABLE 2. DEVELOPED MINES IN NAMIBIA

Uranium mine	Location, region	Annual production capacity, recent production (t U)	Status
Rössing uranium mine	12 km from Arandis (70 km inland from Swakopmund) Erongo region	3820 1790 (2017)	Operational for more than 40 years Longest running uranium mine in Namibia
Husab uranium mine	60 km from Swakopmund Erongo region	5680 1140 (2017)	Started production at the end of 2016
Langer Heinrich uranium mine	80 km east of the major seaport of Walvis Bay Erongo region	2035 1526 (2017)	In production since 2007 Placed in care and maintenance in 2018 owing to low uranium spot prices
Trekkopje uranium mine	35 km north of Arandis Erongo region	2540 373 (2013)	In care and maintenance since 2013

of uranium, as do the Trekkopje and Langer Heinrich mines, which had to be put on care and maintenance in 2013 and 2018, respectively.

Climate change and global warming have in recent years fuelled the world's need for low CO_2 electricity generation, and today more nuclear power plants are under construction than at any other time in recent decades. It is therefore expected that uranium prices will rise. This will create a favourable climate for growth in the uranium industry, which is very important for the socioeconomic development of the Erongo region and of Namibia.

I.1.2. Employment

The mining sector in general is a significant employer in Namibia, a large country with a low population density and a population of only 2.4 million people. The importance of uranium mining for the economy of Namibia, and especially the Erongo region, cannot be overestimated. Uranium mining has created many job opportunities, not only in the mining industry itself, but also

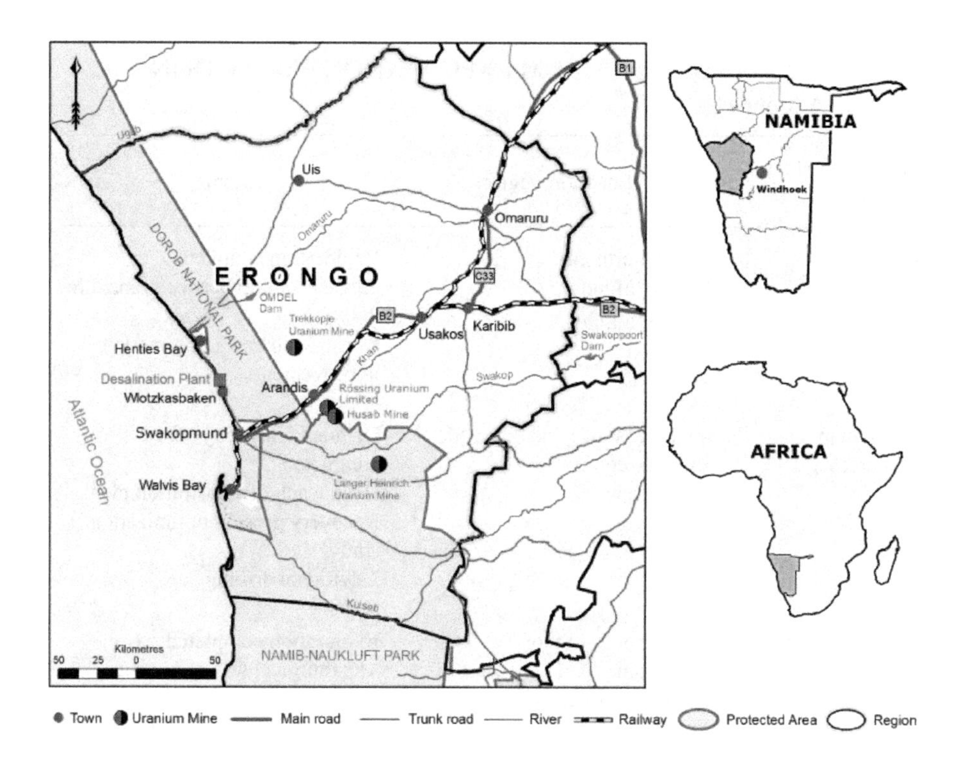

FIG. 4. Map of the Erongo region indicating the location of the uranium deposits in Namibia (reproduced with permission, © Rössing Uranium Limited).

TABLE 3. ADVANCED URANIUM EXPLORATION PROJECTS IN NAMIBIA

Uranium exploration project	Location, region	Status
Zhonge Resources	90 km north-east of Swakopmund, north-east of the Rössing and Husab mines Erongo region	Exploration Metallurgical testing 25 year mining licence granted in 2012 Construction awaiting further exploration and market improvements

TABLE 3. ADVANCED URANIUM EXPLORATION PROJECTS IN NAMIBIA (cont.)

Uranium exploration project	Location, region	Status
Valencia	95 km north-east of Swakopmund Erongo region	Exploration complete 25 year mining licence granted in 2008 Construction awaiting market improvements
Bannerman Resources	48 km east of Swakopmund Erongo region	Detailed feasibility study completed Heap leach demonstration plant Recovery process optimization study Additional drilling
Reptile Uranium	63 km south-east of Swakopmund, 42 km west of Langer Heinrich mine Erongo region	Exploration completed Metallurgical testing in progress
Marenica	87 km north-east of Swakopmund, 30 km north of Rössing mine Erongo region	Exploration completed Patented new metallurgical process

among suppliers and service providers. In 2018, the uranium exploration and mining sector employed some 4400 people. This figure represents around 17% of the total employment in exploration and mining. In addition, on average, every job in the minerals industry generates seven other jobs in the supplier industry. Therefore, the sum of direct and indirect employment emanating from the Namibian uranium sector is assumed to be approximately 35 000 people. Taking into consideration that in Namibia every employed person has approximately five dependents, the sector supports about 175 000 people out of a total population of 2.4 million — a significant proportion.

I.1.3. Contribution to the local and national economies

The Namibian mining sector is the backbone of the Namibian economy. As with employment, the contribution to the local and national economies is

TABLE 4. CONTRIBUTION OF NAMIBIAN MINING INDUSTRY AND URANIUM MINING INDUSTRY TO THE NATIONAL ECONOMY [95, 98]

Year	2013	2014	2015	2016	2017
Contribution of mining to GDP	13.2%	12.2%	11.7%	12.0%	12.2%
Contribution of uranium sector to GDP	1.5%	1.1%	1.1%	1.1%	0.7%

substantial. It accounts for about 50% of Namibia's export earnings. The overall GDP contribution to the economy has ranged between 11% and 12% since 2013, as summarized in Table 4 [95, 98]. In 2017, the uranium sector comprised 0.7% of Namibian GDP.

Namibia has developed two significant uranium mines, Langer Heinrich and Husab, which opened in 2006 and 2016, respectively. Together with the Rössing mine, these facilities contributed 5.3% of the world's uranium mining output in 2017, making Namibia the fourth largest supplier of uranium in the world in 2017. It is expected that Namibia will be an even more significant producer of uranium in the near future owing to current active mine developments and exploration activities. The uranium mining subsector recorded strong growth in terms of real value added of 23.4% during 2017 compared with 13.6% in 2016. This was due to the increase in the production of uranium, mainly through the increase of uranium production at the Husab mine, even though the price of uranium remained low in 2017.

In 2017, NAD[6] 1.5 billion was spent on salaries and NAD 127.2 million was spent on exploration. Government revenue from royalties and levies amounted to NAD 164.3 million. If the uranium price increases, which would make operations profitable, this figure will increase substantially through tax payments. NAD 3.6 billion was spent on fixed investments and NAD 4.84 billion went to local procurement. Indeed, the western Erongo region, where the uranium activities are taking place, presents a different economic picture compared with before the start of uranium mining.

In addition, there are many indirect benefits generated by uranium mining and exploration, such as: the personal taxes of employees; salaries available for spending; employment created; value added tax; corporation tax and personal taxes paid by the service industry; revenue collected and

[6] NAD: Namibian dollar.

jobs created by parastatals (Namwater, Nampower, TransNamib, NamPort); support to the Namibian Institute of Mining and Technology; in-house training; revenue collected and jobs created in other industries (e.g. teachers, doctors, nurses, medical facilities, restaurants, shops); municipal charges; infrastructure development (e.g. desalination, pipelines, powerlines, roads, housing); housing sold to create ownership; and environmental programmes.

I.1.4. Overview of Namibian national policy and government–industry relations

In Namibia, the fundamental principle of the minerals sector is that all mineral rights are vested in the state. Any right granted under any provision of the Minerals (Prospecting and Mining) Act No. 33 of 1992 in relation to the survey or prospecting for, the mining and sale or disposal of, and the exercise of control over any mineral or group of minerals, despite the ownership of any land, belongs to the state.

The Namibian Minerals Policy of 2003 outlines the guiding principles and direction for the industry and the values of the Namibian people in support of the development of the mining sector. It is currently under revision to provide for changes that have taken place in the last 15 years.

Royalties are levied according to the Minerals (Prospecting and Mining) Act as a percentage of the market value of the minerals extracted by licence holders in the course of exploration or mining activities. A rate of 3% is payable on nuclear fuel minerals.

The Export Levy Act became effective on 1 June 2016. The purpose of this act is to impose an export levy on certain goods exported from Namibia to improve the country's value share in its resource base and to encourage further processing and value addition to such goods to support national industrial development.

While Namibia is an IAEA Member State that wishes to enhance existing capacity and capability, the Namibian Government had to impose a moratorium on the issuance of exclusive prospecting licences for nuclear fuel minerals in 2007 owing to an exceptionally high number of applications. The moratorium was in place for ten years and ended on 15 December 2016 to provide an opportunity for further uranium exploration.

In the mid to late 2000s, when the price of fuel for civil nuclear reactors was rising rapidly, resulting in a worldwide boom in uranium exploration and mining, the Namibian uranium industry recommended to the Namibian Government to undertake a strategic environmental assessment of the Namibian uranium province, where exploration for uranium was also expanding rapidly. This assessment was carried out by the Ministry of Mines and Energy, providing vision and generating a culture of cooperation between the uranium mining industry, the Namibian Government and the public.

The Strategic Environmental Management Plan (SEMP) was developed as a result of the strategic environmental assessment. It is an overarching framework and roadmap within which individual projects need to be planned and implemented. It addresses the cumulative impacts of existing and potential developments and the extent to which uranium mining impacts central Namibia. The SEMP has 12 themes, the 'environmental quality objectives', with each articulating a specific goal, providing context, setting standards and having a number of key indicators that are monitored. These themes include socioeconomic development, employment, infrastructure, water, air quality, health, effect on tourism, ecological integrity, education, governance, heritage and future, mine closure and future land use. Annual SEMP reports measure the performance of the 12 environmental quality objectives. The industry is actively contributing to the compilation of annual SEMP reports.

I.1.5. Namibian regulatory bodies

The Namibian Ministry of Mines and Energy regulates the mining industry, including uranium mines. It is responsible for the administration of prospecting, exploration and mining licences and for monitoring the performance of licences with respect to work carried out, production, health and safety, the environment and royalty payments.

The Minerals (Prospecting and Mining) Act 33 of 1992 is the principal legislation for the granting of exploration and mining licences. Other acts relevant to uranium exploration and mining are the Environmental Management Act No. 7 of 2007, which regulates the sustainable management of the environment and the use of natural resources. The Atomic Energy and Radiation Protection Act No. 5 of 2005 provides for protection of the environment and of people in current and future generations against the harmful effects of radiation by controlling and regulating the production, processing, handling, use, holding, storage, transport and disposal of radiation sources and radioactive material, and controlling and regulating prescribed non-ionizing radiation sources.

Reporting is a statutory requirement in terms of the Minerals (Prospecting and Mining) Act. Uranium exploration projects are required to submit quarterly progress reports, while operational mines are obliged to submit reports in a prescribed format, as listed in Table 5.

Namibia enforces a multi-agency approach when conducting inspections at the uranium mines and among the government institutions that conduct inspections and administer different regulatory frameworks and legislation. The Ministries of Mines and Energy, of Health and Social Services and of Labour,

TABLE 5. REPORTING REQUIREMENTS

Report	Frequency
Monthly return	Monthly
Annual return	Annual
Annual financial statements	Annual
Material exported pursuant to 34(a) [37]	Annual
Accident report	Within 36 h of occurrence and annual
Material stock balance report	Every six months

Social Welfare and Employment Creation work closely together to ensure that inspections and enforcement of the various legal provisions and requirements are implemented effectively and efficiently.

Inspections and verification visits to uranium prospecting, exploration and production sites are devoted to visiting critical areas, providing advice and education, observing containment and surveillance measures around the mines, and organizing follow-up visits to verify and ensure compliance. The inspections and verifications also assist licence holders and operators in complying with legislation and regulations and implementing the licensing conditions as per the original authorizations issued by government ministries and institutions. The inspections also include the following activities:

(a) Provide information and support to ensure compliance with the applicable acts and regulations;
(b) Inspect, monitor and analyse data and conduct investigations to measure compliance;
(c) Recommend options for compliance;
(d) Implement actions deemed necessary to prevent or minimize danger to the environment and the public and to prevent the theft and trafficking of radioactive source material.

Other bodies of relevance to uranium exploration and mining are the National Radiation Protection Authority (NRPA) and the Atomic Energy Board of Namibia (AEB). The NRPA serves as the administrator of the Atomic Energy and Radiation Protection Act. Its main duties are to: maintain an inventory and record of activities (i.e. production, processing, handling, transport, use, storage and disposal) involving radiation sources and radioactive and nuclear material in Namibia; regulate all activities involving radiation sources and radioactive and nuclear material in Namibia; inform the AEB about the extent of radiation exposure; and enforce all provisions of the Atomic Energy and Radiation Protection Act.

The NRPA was established as an independent regulatory body. It needs to act independently in the exercise of functions under the Act and considers only the relevant provisions of the Act and scientific and technical matters that may be relevant to the issue concerned. The NRPA is not a juristic unit and therefore does not have administrative autonomy as is the case with other state enterprises. The organizational requirements for the NRPA and for the staff performing the work of the AEB are contained in the Directorate Atomic Energy and Radiation Protection Authority, an administrative entity in the Ministry of Health and Social Services. The NRPA functions independently regarding technical and scientific matters within the scope of the Act, but functions administratively as a directorate in the Ministry of Health and Social Services.

The AEB was established in 2009 pursuant to the requirements under Section 3 of the Atomic Energy and Radiation Protection Act 2005. The AEB is an advisory body reporting to the Minister of Health and Social Services. The secretary of the AEB is the director of the NRPA, which is the technical arm responsible for the administration of the Act. The AEB manages Namibia's nuclear and radioactive materials in a manner that safeguards people and respects and protects the environment today and in the future. By providing appropriate advice, the AEB ensures that the use of radiation and nuclear energy in Namibia does not cause unacceptable impacts on the health of workers, members of the public and the environment.

I.1.6. International commitments

Namibia is party to several international agreements for nuclear safety, security and safeguards, and recognizes international principles and standards. It has committed to adopt the highest levels of industry performance to regulate its uranium industry. The country has established the appropriate regulation for its uranium industry owing to the nature of the uranium sector and the resulting level of public interest. As a signatory to the Treaty on the Non-Proliferation of Nuclear Weapons, Namibia also implements its obligations under the treaty.

The Namibian Government also ensures that all industry stakeholders are aware of the country's international reporting requirements as per IAEA guidelines and standards through regular information sharing workshops. It is important for Member States to provide accurate and timely data to the IAEA to enable it to draw safeguards conclusions and provide assurance that nuclear materials in its possession are properly classified and accounted for. Namibia is committed to providing the IAEA with accurate and timely information concerning nuclear material subject to safeguards and the description of facilities relevant to safeguarding such material.

I.1.7. Overview of the Namibian Uranium Association and the Namibian Uranium Institute

The Namibian Uranium Association (NUA) believes that the advantages of uranium make it an appropriate source of energy for the 21st century, especially because of its low carbon footprint. NUA members are the uranium producers, the majority of uranium exploration companies active in Namibia, contractors, suppliers and service providers.

The Namibian Uranium Institute (NUI) was established by the NUA as part of the stewardship mission. NUA's Scientific Committee, comprising respected and independent scientists, oversees the work of the NUI. The latter serves as a communication hub for the Namibian uranium industry and its suppliers and service providers. The institute offers training in radiation safety, health, environmental management and occupational hygiene. Through identification of best practices and provision of information about their implementation, it gives members an opportunity to collectively advance safety and health performance. NUI cooperates with Namibian ministries and state agencies, as well as with the Namibian University of Science and Technology.

In 2013, the directors of the NUA established the Sustainable Development Committee, which aims to ensure that uranium for the nuclear fuel cycle produced in Namibia is explored for, produced, transported, stored and managed in a socially, economically and environmentally responsible manner. The committee identifies risks and has put in place a number of supporting technical working groups that address and advise on emerging issues. It also advises the NUA on policy issues, especially with reference to sustainable development.

Special emphasis is given in this context to public participation, intergenerational equity, sustainable use of natural resources and public access to information. Risk assessment and monitoring with reference to health, environment, radiation safety and security are other duties of the Sustainable Development Committee. It is also involved in the advancement of internal compliance and control systems, in measures to manage risks, in the assessment of the efficiency of controls in place and in making recommendations to the NUA concerning risk management.

The International Council on Mining and Metals provides guidance to the Namibian uranium industry on the sustainable development of the mining and metals sector. This means that investments need to be technically correct, environmentally sensible, commercially viable and socially acceptable. Standards for operation that guarantee a good international standing and reputation are set, as they are critical for any mining company in obtaining and maintaining a 'social licence to operate' in a given community. It is indispensable to address environmental, economic and social aspects through all stages of mineral projects, from exploration, construction and operation to mine closure. The Sustainable Development Committee plays an important role in maintaining these standards.

I.2. MILESTONES ASSOCIATED WITH NAMIBIAN URANIUM PROJECTS

I.2.1. Rössing mine

Rössing Uranium, which is majority owned (68.62%) by China National Uranium Corporation Ltd (CNUC), is one of Namibia's two currently producing uranium mines. In 2019, Rio Tinto sold its controlling stake to CNUC. The Rössing mine is co-owned by the Iranian Foreign Investment Company, the Industrial Development Bank of South Africa, the Namibian Government (which in addition has 51% voting rights) and a number of private shareholders.

At the Rössing mine, uranium mineralization is hosted by early Paleozoic alaskites. Mineralization occurs in veins and disseminated grains within and adjacent to the alaskites. On 1 January 2017, recoverable resources amounted to 77 956 t U at an average grade of 0.025% U [2]. The Rössing mine is the world's longest running open pit uranium mine and celebrated 40 years of operation in 2016. It has a nameplate capacity of 3820 t U per year and by 2017 had contributed a total of 112 450 t U to the global nuclear fuel cycle.

The mine is situated in Namibia's Erongo region, about 12 km from the town of Arandis, which is some 70 km north-east of the coastal town of Swakopmund. The mining licence and accessory works areas measure around 180 km^2, but only 25 km^2 of the total area is used for mining, processing and overburden and tailings storage facilities. The mine is an open pit operation with conventional blasting, loading and hauling. The open pit currently measures approximately 3 km × 1.5 km and is ~390 m deep.

I.2.1.1. Stakeholder involvement

Rössing Uranium's stakeholders include its shareholders, employees and contractors, the communities of Arandis, Swakopmund and Walvis Bay, Namibian Government institutions, service providers and the mine's customers. Regular information sharing with both internal and external stakeholders is a key enabler of the company's business success. The company's corporate communication section uses various platforms, initiatives and activities to establish, nurture and maintain good relationships and promote the sharing of information with all stakeholders, as well as to receive concerns, if any. For internal stakeholders this includes newsletters, the intranet and various written and verbal communications, while external stakeholders are engaged through the annual publication of a stakeholder report, a website, a visitor's programme and regular communication with key stakeholders and the media. Furthermore, the company supports the Rössing Foundation, which is involved in many community and educational activities, such as a mobile laboratory in support of educational programmes.

I.2.1.2. Safety and radiation protection

Rössing Uranium is committed to zero harm and has put in place rigorous processes to ensure the safety of every employee and contractor. The identification and management of material risks are crucial to its business approach, and a formalized, integrative HSE management system is utilized to optimize, coordinate and manage the operations, personnel, plant and equipment.

The structure of the HSE management system generally follows the layout of common international standards such as ISO 14001, and the Occupational Health and Safety Assessment Series of British Standard 18001. An auditing programme periodically evaluates the effectiveness of the HSE management system.

Rössing Uranium's radiation management plan provides a comprehensive summary of the risk assessments, sources and receptors, and the controls implemented. It is updated annually. The NRPA audits the implementation of the radiation management plan. Workers who are considered to be at an elevated risk for radiation exposure — namely anyone who is at risk of receiving an annual dose of 5 mSv or more from all exposure pathways combined — are termed 'radiation workers'. Such workers receive continuous gamma monitoring in the form of a dosimeter and undergo monthly urine testing to check for accidental ingestion of uranium. Female radiation workers must undergo monthly pregnancy tests so they can be protected from any exposure in the case of a pregnancy.

Figure 5 illustrates the personal radiation exposure dose for various similar exposure groups in relation to the regulatory annual dose limit of 20 mSv.

I.2.1.3. Environmental protection

Rössing Uranium is committed to protecting the environment in which the company operates. Measures include a wide range of preventive monitoring activities, with a particular focus on water management and monitoring,

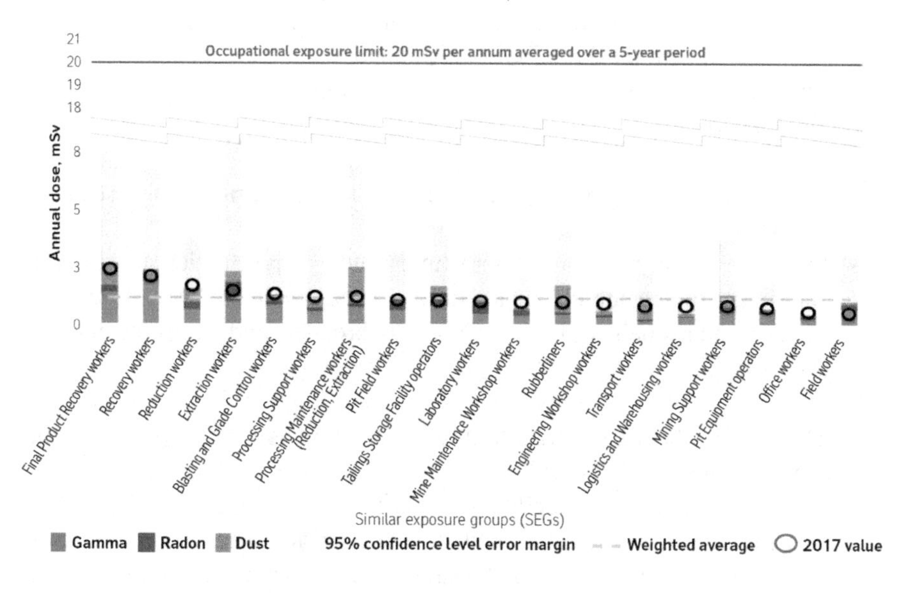

FIG. 5. Annual personal worker radiation exposure for Rössing Uranium mine workers.

especially in the light of the extremely low rainfall associated with the Erongo region's water scarce, hyper-arid climate. The company has a strong history of engagement and cooperation with regulators and other stakeholders to ensure that the environment remains protected.

Environmental impacts are managed with guidance from, among others, Namibian legislation, the ISO 14001 environmental management system, Rio Tinto's performance standards and international best practices. Transparent reporting assures stakeholders that the company's environmental impacts are monitored and the necessary mitigation measures are in place to keep these impacts ALARA.

I.2.1.4. Protection/enhancement of cultural, tourism, farming, pastoral and similar interests

The Rössing Foundation was established in 1978 through a deed of trust as a vehicle to oversee and implement many of Rössing Uranium's corporate social responsibility activities in Namibia. The activities that the Rössing Foundation drives or supports are formulated in a memorandum of understanding between the foundation and partner organizations, but in particular the seven education directorates. These critical partners include the Ministry of Education, Arts and

Culture; Ministry of Mines and Energy; the National Institute for Educational Development; UNICEF (United Nation's Children Emergency Fund); Erongo Regional Council; and the Arandis Town Council.

The Rössing Foundation focuses mainly on programmes that target the following:

- — Teacher and pupil support programmes to improve primary and secondary education;
- — Provision of bursaries and part-time study opportunities to deserving Namibians to develop the local workforce and specialized vocational skills;
- — Medium and small scale enterprises to broaden and reinforce the local economy.

The diverse board of the Rössing Foundation is made up of members with a wide range of skills and experience. The Rössing Foundation and the Rössing Environmental Rehabilitation Fund are managed independent of Rössing Uranium by their trustees and were created for special purposes. Some trustees also serve on the board of Rössing Uranium, which has a unitary board with the roles of chairperson and managing director separate and distinct.

I.2.1.5. Funding and financing

The responsibility for monitoring and sanctioning the financial statements to ensure that they properly represent the company's business and its profits and losses at the end of each financial year lies with the directors. An opinion as to whether the financial statements fairly represent the company's financial position is obtained from independent auditors.

In preparing the financial statements, Rössing Uranium's management adheres to the International Financial Reporting Standards and the Namibian Companies Act (Act No. 28 of 2004, amended in 2011).

I.2.1.6. Safeguards and security

Namibia is a signatory to the relevant international instruments for safeguards and security of nuclear material, which also cover the operations of Rössing Uranium. Moreover, the company regards international best practices and product stewardship as a foundation for conducting business. Regular reporting to the Ministry of Mines and Energy is carried out and regular IAEA inspections take place.

I.2.1.7. Transport/export route

Uranium oxide is transported to the company's customers in 205 L drums that conform to IAEA standards. To complete the packaging process, the external surface of each drum is cleaned and decontaminated and a contamination swipe is taken to ensure that the drum meets shipping requirements. The drum is then labelled with the required transport information on two sides. The drums are loaded into a shipping container labelled on all four sides with labels that are required for shipment of UOC.

The consignment leaves the Rössing mine by rail to the port at Walvis Bay — a journey of about 90 km. Walvis Bay harbour has an area designated for uranium exports, as well as the safeguards instruments required for the temporary storage and handling of radioactive material. The consignment is then loaded onto a ship and transported in accordance with the approved safeguards and security measures to an overseas port for onward transport to converters.

I.2.1.8. Human resources development

Capacity building at Rössing Uranium is a critical process aimed at enhancing productivity and organizational performance. The company's training and development section supports the mine's strategy to achieve its objectives by providing support and services to the various departments through collaboration and partnerships. Various initiatives, such as bursaries, employee engagement and apprentice job attachments, are implemented to achieve the goal of empowering and developing the workforce.

Technical training is essential to ensure that the knowledge, skills and attributes of the workforce are enhanced. Various training interventions are therefore conducted to enhance skills upgrading, efficiency and effectiveness. Rössing Uranium also contributes to the Namibia Training Authority's Vocational Education and Training Fund.

I.2.1.9. Site and supporting facilities

The site is connected to the Namibian road infrastructure through a short access road joining the Trans-Kalahari Highway, which passes the mine in the north. The highway links the harbour of Walvis Bay with Namibia's neighbours to the east and north-east. The site is also connected to the Namibian railway grid, which allows delivery of goods and dispatching of the final product by train. For its water supply, the Rössing mine is connected via a 70 km long pipeline to the state operated Central Namib Water Scheme. The national electricity grid, which passes directly by the mine, is connected by a substation. The mine has town offices in Swakopmund and Windhoek.

I.2.1.10. Contingency planning

Under the structure of Rössing Uranium's HSE management system, contingency plans are in place for all areas and processes requiring such planning.

I.2.1.11. Waste management and minimization

Rössing Uranium's waste management system distinguishes between uranium mineralized waste and non-mineralized waste. Mineralized waste is defined as including mine rock and overburden, tailings and spent heap leach ore from mineral processing. At the Rössing mine, the mineralized waste currently identified is mine rock, overburden and tailings. All ex-pit material below 100 ppm is classified as waste and is disposed of as such. The waste dumps are not separated by rocks or any other geological property. The mineralized waste management system aims to: decrease and manage risk from operations in terms of human health and the environment; identify and assess risks (worst case and normal operating conditions); implement measures to control and manage negative impacts of mineral waste disposal; and monitor pollutants to ensure that Rio Tinto standards and international standards are met.

The tailings storage facility to the north-west of the plant and mine was originally designed as an upstream ring deposition facility in a gorge and was operated as such until the early 1980s. By that time, surface seepage of tailings liquid created another gorge towards the west, and a modified ring deposition layout was implemented, confining deposition to the catchment of the original gorge, which is protected by a surface seepage collection dam situated in the main channel of the gorge, about 1 km downstream of the facility. It is anchored on its eastern end against a north-east trending ridge of hills. Today, the tailings storage facility is the largest feature on the Rössing site, covering a footprint of about 730 ha. It rises to an elevation of about 100 m above the surrounding surface and is one of the largest uranium tailings facilities in the world.

All tailings from the uranium extraction process are conveyed and pumped to the tailings storage facility. Owing to the low uranium content of the ore, the tailings material, which is quite coarse, consists of virtually the entire mass of input ore plus waste process liquids. Surface seepage from the tailings impoundment occurs through a filter drain in the embankment and the foundation materials. An extensive seepage control programme and monitoring system has been established to contain subsurface seepage. To reduce the wetted surface area, tailings discharge at any point in time is confined to 40 ha paddocks, with only one paddock surface wetted during tailings discharge, thereby reducing the wetted area by about 90%.

A non-mineralized waste management plan is in place to ensure that regulatory and internal requirements have been addressed and to minimize the environmental, safety and health hazards associated with the handling, storage and disposal of the variety of waste products generated by activities, products and services at the Rössing mine. The waste management hierarchy of eliminate, reduce, reuse, recycle and dispose is followed throughout.

The non-mineralized waste management plan outlines how and what Rössing Uranium is doing, and has done, to reduce the amount of pollution and the generation of waste. Goals for pollution prevention are measured by compliance with the established operations. When new waste types are generated, new disposal options will be researched and included in the plan. The document is updated as the need arises. The procedure identifies all waste streams, indicates disposal requirements and outlines record keeping requirements.

At present, non-mineral waste materials include wastewater not generated from the mineral ore, scrap materials, redundant conveyor belts, domestic waste, and used oils and lubricants from maintenance activities. A waste contractor handles recyclable waste materials such as scrap metal, wooden pallets, paper, plastic and metal containers on-site. Rössing Uranium also operates a bioremediation facility for oil sludge soil. Through an aggressive wastewater recycling programme, the Rössing mine continuously reduces its freshwater requirements, thereby minimizing wastewater.

I.2.1.12. Industrial involvement, including procurement

Rössing Uranium is a Class A Founder Member of the Chamber of Mines of Namibia, and a founding member of the NUA. The company makes substantial contributions to the activities of these two bodies. It is also a substantial contributor to the economic development of the Erongo region, and of Namibia, as it is a major employer and purchaser of goods and services. The mine's annual procurement expenditures have a significant 'multiplier effect' — the phenomenon where spending by one company creates income for further spending by others. In 2017, procurement of goods and services for the mine's operations amounted to NAD 2.3 billion, of which 73.5% was spent with Namibian suppliers.

I.2.2. Husab mine

Swakop Uranium, the owner of the Husab mine, is a partnership between Namibia and China, of which 10% is held by the Namibian State owned

Epangelo Mining Company and 90% by Taurus Minerals Ltd. Taurus Minerals Ltd in turn is jointly owned by China General Nuclear Power Group and China Africa Development Fund. The partnership culminated in China's single largest investment in Africa, the construction of the world class Husab mine. Swakop Uranium is a private company registered in Namibia.

The discovery of the Husab uranium deposit in February 2008 was one of the world's most significant uranium mineral finds in decades. The Husab mine has an alaskite hosted deposit, similar to that of the Rössing mine. The deposit lies under a shallow alluvial sand cover. On 1 January 2017, recoverable resources amounted to 187 546 t U, at an average grade of 0.033% U [2]. In 2011, Swakop Uranium was granted a licence to develop the Husab mine, which is set to become the third largest uranium mine in the world. Mine development commenced in 2014. The total investment for the Husab mine is about $5.2 billion, and more than $2 billion was required to build the mine. The mine will more than double current Namibian uranium production and propel Namibia into third place in terms of global uranium production. The mine produced the first drum of uranium for the export market in December 2016, a significant event in the history of Namibia.

The mine is located about 5 km south of the Rössing mine and 45 km north-east of Walvis Bay, Namibia's only deep water harbour. The mine site encompasses mining licence and accessory works areas of about 110 km^2, of which about one third is used for mining, waste disposal and processing. The tailings storage facility alone covers about 5 km^2. Mining is performed by blasting, loading and hauling from the open pit before the uranium-bearing rock is processed to produce uranium oxide.

Swakop Uranium supports and promotes the Government of Namibia's National Agenda and Harambee Prosperity Plan. The company is the largest employer in the Namibian mining industry, with 1650 permanent employees and about 500 contractor employees, thus assisting the Government to reduce unemployment and alleviate poverty.

I.2.2.1. Stakeholder involvement

Swakop Uranium's stakeholders include: the Government as regulator; shareholders; employees; contractors; the communities of Arandis, Swakopmund and Walvis Bay; Namibian Government institutions, in particular the Directorate of Parks and Wildlife as the custodian of the Namib-Naukluft Park, where the Husab mine is located; the conservation and scientific community; the service providers; and the mine's customers. To share information, regular visits of stakeholder groups are facilitated, and the work of the NUA, the NUI and the Chamber of Mines of Namibia is actively supported.

Employees are a very important stakeholder group, and Swakop Uranium and the mine Workers Union of Namibia signed a historic three year wage agreement in 2016 to regulate the conditions of the employment framework, including a housing allowance for employees in the bargaining unit to purchase or rent accommodation units. The two parties have a cordial and constructive relationship, aligned with the vision of the company to work through its STARIC (safety, transparency, accountability, respect, integrity, collaboration) values to build the Husab mine into a world class operation.

I.2.2.2. Safety and radiation protection

Swakop Uranium's HSE management requires employees to be safety conscious, encourage their fellow workers to work safely and diligently, and reduce incidents to zero base, as negligent behaviour leads to injuries and damage to company property. Employees must also: check machines and equipment or company assets before the start of work; report any defects immediately to the supervisor; work in teams and share work experience and any shift issues with colleagues to avoid making the same mistakes; focus on quality of work; and perform tasks correctly to avoid errors and repetition.

Swakop Uranium has established a radiation management plan that is regularly audited to ensure compliance. A network of dust fallout measuring stations ensures the prevention of inhalation of radioactive material, and voluntary urine testing is available for the employees. Workers considered to be at risk of receiving an annual dose of 5 mSv or more from all exposure pathways combined receive continuous gamma monitoring in the form of a dosimeter. Environmental radiation monitoring consists of a soil sampling and radionuclides analysis baseline study completed at the end of 2016, annual monitoring for and analysis of aquatic and atmospheric radionuclides, and a thorium and uranium survey.

Swakop Uranium complies fully with the provisions of the Namibian Atomic Energy and Radiation Protection Act (No. 5 of 2005) and its regulations, and reports to the NRPA, which has approved Swakop Uranium's radiation management plan, on a regular basis. Swakop Uranium also complies with all guidelines, standards and provisions of the IAEA, of which Namibia is a member.

I.2.2.3. Environmental protection

Swakop Uranium has an environmental management plan committed to caring for all species of fauna and flora found near or within its exploration and mining areas. Its environmental department provides guidance and advice on new projects and activities, as well as environmental management plan commitments and legal requirements, and is guided by legislation, best practices, relevant

EIAs and scoping reports, and the operational Husab environmental management plan. The department conducts compliance monitoring by means of inspections and audits in the various work areas, implements and conducts environmental monitoring and baseline establishment, and acts as the link between the authorities and Swakop Uranium management on environmental issues. It is the custodian of environmental permits and licences and is responsible for biannual reporting to authorities and external annual environmental management plan audits.

Swakop Uranium has supported substantial research on *Welwitschia mirabilis*, Namibia's ancient national plant, which grows in areas around the mine. Carbon dating shows that medium sized plants can be 1000 years old.

The Husab mine also needs to ensure that limited nearby water resources are not adversely affected by mining operations. Long term records from the Rössing mine (situated 5 km north of the Husab mine) show an annual average rainfall of between 30 mm and 35 mm. A hydrogeology report commissioned by Swakop Uranium concluded that the mining activities will have an effect on water levels. Although there are no settlements in the area, Swakop Uranium has drilled a number of groundwater monitoring holes around the pit, the mine rock dump, the tailings storage facility, the *Welwitschia mirabilis* fields, and the Khan and Swakop rivers to determine the effect of mining activities in the area. Water levels in all boreholes are measured monthly and strategic boreholes are sampled every three months for water quality analysis by a third party (SLR Environmental Consulting). However, production boreholes have not been in operation, as the company has moved from construction to a mining operation and currently obtains its water from the Orano desalination plant. Finally, Swakop Uranium complies with all applicable international standards, as adopted for the Husab mine operational requirements.

I.2.2.4. Protection/enhancement of cultural, tourism, farming, pastoral and similar interests

The Husab mine lies within one of Namibia's national parks, the Namib-Naukluft Park. Interaction with the conservation and tourism stakeholders therefore occurs on a regular basis, and some tourist routes are maintained by the mine. Furthermore, limited irrigation farming takes place in the Khan and Swakop rivers adjacent to the mine, and regular contact is therefore maintained with the farmers to ensure that there are no adverse effects of the mining and processing activities on the quality of the groundwater.

A heritage and archaeological chance find management programme is in place to ensure the safety of any item or structure protected under Namibia's National Heritage Act.

As Swakop Uranium is committed to social and empowerment issues, the Swakop Uranium Foundation engages poor and vulnerable communities living close to the park's borders to address critical needs, create a better future and ensure their growth together with the mine's.

I.2.2.5. Management coordination/facilitation

Swakop Uranium's board of directors provides strategic guidance, as well as a compliance and auditing function. The executive committee of the mine is tasked with executing board directives to build an inclusive and high performance culture for all employees through the establishment of well structured departments with clear accountability levels. All employees are trained to understand their role in the company.

I.2.2.6. Funding and financing

The shareholders of the company were responsible for obtaining the necessary finance to construct the Husab mine. During ramp-up operations, they have continued to ensure that the mine operates with a positive cash flow for its continued existence.

I.2.2.7. Safeguards and security

Namibia is a signatory to the relevant international instruments for safeguards and security of nuclear material, which also cover the operations of Swakop Uranium. Regular reporting to the Ministry of Mines and Energy is carried out and regular NRPA inspections take place.

I.2.2.8. Transport/export route

Uranium oxide is transported to its destination in 210 L drums. The cleaned drums are loaded into steel shipping containers. Consignments leave the Husab mine by road to the port of Walvis Bay, which is about 90 km away. Further onward transport is by ship, in accordance with safeguards and security provisions.

I.2.2.9. Human resources development

Swakop Uranium is committed to the continuous improvement of its workforce through both formal and on the job training. Bursaries are provided for young Namibians to realize their tertiary education goals. Job attachments for students are facilitated through memoranda of understanding with the Namibian Institute of Mining and Technology, the Namibian University of Science and Technology and the University of Namibia. The company is focusing in particular on upgrading skills at all levels in the organization and promoting cross-training to ensure that its employees are multi-skilled.

I.2.2.10. Site and supporting facilities (infrastructure)

The site is connected to the Namibian road infrastructure through a 22 km long access road joining the Trans-Kalahari Highway, which passes the mine to the north and links the harbour of Walvis Bay with Namibia's neighbours to the east and north-east. The access road includes a 160 m long bridge, which is the longest bridge constructed in Namibia since independence in 1990.

Water is provided through a 65 km long purpose-built pipeline that links the mine with the Swakopmund water reservoir. This reservoir is supplied by the Omdel Dam scheme, as well as the Erongo desalination plant. The Husab mine is connected to the Namibian electricity grid through the 50 MV·A Lithops substation. The site also produces a smaller amount of electricity from waste heat at its sulphuric acid plant. Swakop Uranium has offices in Swakopmund, as well as in the capital city of Windhoek.

I.2.2.11. Contingency planning

Under Swakop Uranium's HSE management system, contingency plans are in place for all areas and processes requiring such planning.

I.2.2.12. Waste management and minimization

Swakop Uranium's waste management plan focuses on separation and recycling. While hazardous waste obviously needs to be sent to a hazardous waste disposal site, suitable hydrocarbon waste is reused before it is deposited at such a site. Radioactive contaminated material is kept on-site at the tailings storage facility, while rubble, tyres and wood are reused, when applicable. Glass, plastics, metals, cans, cardboard and paper are transferred to a recycling facility. Only general domestic waste goes to landfill. Water is recycled to minimize wastewater and effluents that require management.

While the initial plan was to combine the mine rock facility and the tailings facility, this proved to be impractical, and the mine therefore has two different facilities. At the mine rock dump, the entire footprint is covered with a 2 m thick layer of calcretic overburden soils to maximize the potential for neutralization of any acid seepage that might occur. This layer has a very high neutralizing capacity and thus maximizes the potential for neutralization of seepage. Surface water in the form of runoff from the slopes is captured by a dirty water collection channel and directed to a storage pond for subsequent use as dust control or within the plant.

The tailings storage facility site was selected to minimize impacts on the ephemeral runoff channels and near surface aquifer systems, and hence is in a location outside the zone of ephemeral channels. Furthermore, it was placed at the lowest elevation possible to minimize the pumping costs and risks associated with high pressure pipelines. The tailings storage facility was designed and operates as an upstream raise, whereby a relatively low height starter wall is constructed, and additional volume is created by raising the walls with an outer layer of cover soil underlain by tailings. It is lined with a geomembrane liner to minimize seepage losses, both for maximizing water return and for minimizing pollution potential. To facilitate return of water from the dam, an actively managed decant pond is in place.

Swakop Uranium has a contract with a professional waste management organization that provides on-site services, such as provision of the necessary waste management equipment, removal of general waste from the site, hazardous waste removal from the site to Walvis Bay, on-site recycling, on-site spillage cleanups, management of the building rubble yard, on-site medical waste and fat trap management, on-site KleenBin services and provision of on-site HSE staff.

I.2.2.13. Industrial involvement, including procurement

Swakop Uranium is a Class A Member of the Namibian Chamber of Mines and a founding member of the NUA. The company makes substantial contributions to the activities of these two bodies. According to general practice in the Namibian mining industry, procurement is performed locally as much as possible. In 2017, Swakop Uranium spent NAD 2.5 billion in local procurement, a figure that has a very significant impact on the local economy.

I.2.3. Langer Heinrich mine

The Langer Heinrich uranium mine is owned by Paladin Energy Ltd from Australia (51%) and China National Nuclear Corporation (49%). It is a surficial calcrete type deposit associated with sediments occurring within a tertiary paleodrainage system. Uranium mineralization occurs as carnotite containing uranium and vanadium. On 1 January 2017, recoverable resources amounted to 37 623 t U at an average grade of 0.045% U [2]. Between 2007 and 2018, the mine operated a conventional open pit uranium mine located in the Namibia–Naukluft National Park, about 90 km east of Walvis Bay. Uranium production for the period July 2017 to June 2018 comprised 1145 t. Owing to the current low price of uranium, a board decision was taken to prepare the mine for care and maintenance, which was put into effect in May 2018.

The key focus at the Langer Heinrich mine during care and maintenance is the safety of personnel and the security of the project assets. Care and maintenance activities include maintaining the processing plant and equipment in a state of readiness to facilitate a restart of operations, complying with legal and social obligations, conducting environmental and radiological monitoring, and managing tailings facility water.

I.2.3.1. Stakeholder involvement

The Langer Heinrich uranium mine considers stakeholder engagement to be the basis for building strong, constructive and responsive relationships that are essential for the successful management of the mine's environmental and social impacts. Such engagement takes the form of stakeholder mapping and subsequent meetings with interested and affected parties. There is a dedicated email address for queries and concerns, and a formal engagement plan is followed for interaction with internal stakeholders by both management and the mother company in Australia. Regular interaction with the relevant government ministries also takes place, and during the operational phase Langer Heinrich issued a newsletter.

I.2.3.2. Safety and radiation protection

Through its robust health, safety and radiation policies, the Langer Heinrich uranium mine ensures that employees work in a safe environment and with the aim of zero harm. The mine operates within the relevant national and international legal, as well as voluntary internal (e.g. not prescribed by law), requirements. It continuously assesses and reduces risks throughout operation and raises health and safety awareness through specialized training and employee health awareness programmes. The mine's occupational health and safety management system is based on the NOSA CMB253N standard, in accordance with OHSAS 18001.

The radiation management system is implemented in accordance with Namibian national legislation, as well as the fundamental principles of the International Commission on Radiological Protection and related IAEA Safety Standards Series publications (notably GSR Part 3), and is set out in its Safety, Health, Environment and Radiation Protection Policy, Radiation Management Plan and Radiation Standards. It ensures that radiation protection principles are firmly established, that the radiation exposure of all employees and affected persons is lower than the legislated limits and ALARA, and that there are no adverse effects on the regional communities or their environment. The monitoring programme, which forms an integral part of the radiation management programme, was revised to align with the care and maintenance activities. The 2019 monitoring programme results have been used to derive the annual dose assessment for each worker. The total dose is the sum of the individual doses from the following three exposure pathways:

(a) Inhalation of long lived radioactive dust;
(b) Inhalation of radon decay products;
(c) External gamma radiation.

Monitoring is carried out on a statistical sampling plan across similar exposure groups on a risk prioritized basis. The routine doses received by each similar exposure group from the identified exposure pathways in 2019 are presented in Fig. 6.

I.2.3.3. Environmental protection

The Langer Heinrich uranium mine operates under the approved ML 140 and ML 172 Environmental Clearance Certificate conditions [90]. EIAs were conducted during the project development phase in 2005. Further EIAs were undertaken for the Stage 3 and Stage 4 expansion projects in 2010 and 2012, respectively. All of the assessment processes involved extensive stakeholder

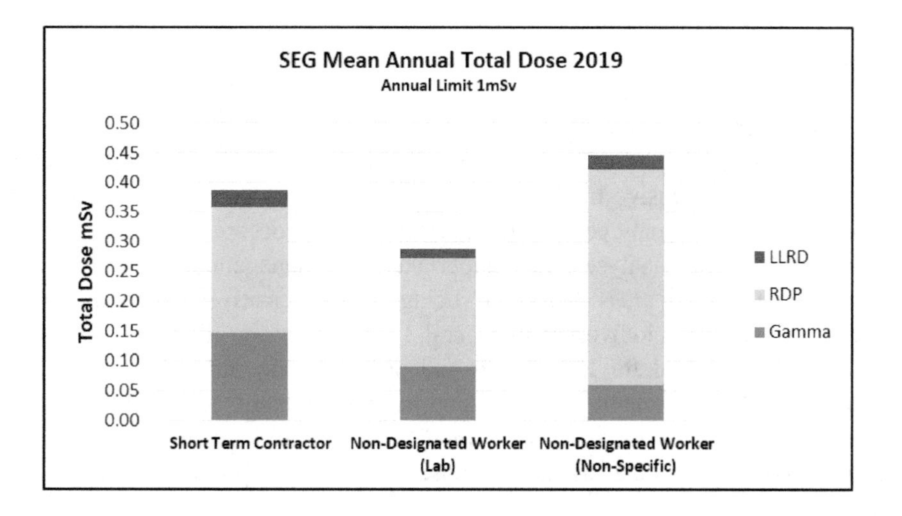

FIG. 6. Annual radiation doses for similar exposure groups (SEGs) among Langer Heinrich mine employees in 2019. LLDR: long lived radioactive dust; RDP: radon decay products.

consultation, and the reports were made available for public review and comment. Prior to project development and expansion projects, environmental baseline studies were conducted, potential impacts were assessed and environmental management plans and monitoring programmes were established to minimize impacts over the life of the mine. The Langer Heinrich mine has an operational environmental management plan that has been submitted, as required, to the Namibian Government and relevant third parties. This environmental management plan consists of 15 management and mitigation plans developed and implemented in accordance with ISO 14001:2004 and South African National Standard 14001:2005 requirements.

The mine follows a water use and quality standard to ensure efficient, safe and sustainable use of water and the protection of water resources and ecosystems around its sites. It undertakes its operations in a way that maximizes the recycling and reuse of water, has a water management and mitigation plan in place, and maintains a comprehensive site water balance to ensure that the company achieves its water usage, supply and resource protection objectives.

Air quality is managed through a formal air quality management and mitigation plan. The mine remains committed to avoiding, preventing and mitigating adverse impacts to air quality generated as a result of any activities. An air quality sampling and monitoring network provides the basis for the mine's air quality management plan. During routine operations, various dust suppression and control measures are applied in the process plant and at work areas with high levels of dust. These procedures will be reactivated once the mine goes back into production.

Prior to the mine's project development and expansion projects, biodiversity baseline studies were conducted and potential impacts were assessed, from which a biodiversity management and mitigation plan was implemented to manage potential impacts, for example in areas where animal and plant species were identified as needing protection. As the mine is located in the Namib-Naukluft National Park, extensive biodiversity studies have been conducted to establish and monitor biodiversity composition, structure and processes. Using the results from these studies, analyses were undertaken and management measures were developed to avoid areas ranked as being highly sensitive and to minimize negative impacts on biodiversity in general. In the area of the mine, approximately 936 ha of land used for the mining and processing facilities is classified as 'disturbed'. Approximately 39 ha of land has been rehabilitated to date (2021). The mine continues to maintain a biodiversity database with historical and current biodiversity data, including conservation status, preferred habitats and recorded sightings.

I.2.3.4. *Protection/enhancement of cultural, tourism, farming, pastoral and similar interests*

There are no communities living permanently in the direct vicinity of the mine that could be directly affected by the mining operation. However, engagement continues with conservation and tourism stakeholders through a formal social performance management plan.

The Langer Heinrich uranium mine supports the Gobabeb Training and Research Internship Programme, which targets young Namibian scientists interested in the fields of conservation and ecological restoration. Under the guidance of researchers from the Gobabeb Research and Training Centre, students design and implement independent research projects that contribute to Namibia's ability to manage and restore degraded ecosystems.

I.2.3.5. *Funding and financing*

The mine is owned and financed by its shareholders. Under operational conditions it is funded from operational cash flow, while in care and maintenance funding is provided by its owners and operator.

I.2.3.6. *Safeguards and security*

Namibia is a signatory to all relevant international instruments for safeguards and the security of nuclear material, according to which the operations of the Langer Heinrich uranium mine are managed and regulated. The company regards international best practice and product stewardship as a foundation for conducting business. Regular reporting to the Ministry of Mines and Energy is carried out and NRPA inspections take place on a regular basis.

Reporting to other relevant national regulatory authorities, such as the Ministry of Environment and Tourism, Department of Water Affairs, NRPA, Ministry of Health and Social Services, and Ministry of Labour, Industrial Relations and Employment Creation, is done against agreed time frames and conditions.

I.2.3.7. *Transportation/export route*

During operation, the Langer Heinrich uranium mine adhered strictly to the terms of the licences covering both the export of the product to various conversion facilities and the transport of metallurgical and geological samples to various laboratories worldwide. In all circumstances, packaging, marking, labelling and documentation for the transport were carried out in accordance with the regulations in SSR-6 [85]. The monitoring of the packages was conducted by competent personnel.

For shipment, the yellow cake product was packaged in 205 L steel drums classified as industrial packaging category 1 (IP-1). Each drum was labelled with a category II label according to the regulations, describing its content as low specific activity category 1 (LSA-1) material. The drums were sealed with steel rings, stacked and strapped securely inside approved ISO freight containers. The packaging, labelling and documentation for each shipment fully complied with SSR-6 [85].

Occasionally the mine ships geological or metallurgical samples to independent laboratories for testing. As this material may contain radioactive material, each shipment is monitored to ensure full compliance with SSR-6 in terms of packaging (UN 2910 excepted packages), labelling and documentation [85].

I.2.3.8. Human resources development

The Langer Heinrich uranium mine has routinely supported a range of educational and skills development initiatives. Through the mine's study assistance programme, the company assisted employees in furthering their educational qualifications for both intermediate and long term development in alignment with business needs. A bursary programme provided financial support to a number of students to pursue formal qualifications in specific fields that were scarce in Namibia and of direct importance to the operations. The graduate development programme focused on attracting graduates and trainees to develop a pool of future skilled individuals and potential leaders. The apprentice programme provided students from the Namibian Institute of Mining and Technology with opportunities to acquire hands-on training in various vocational trades. An in-house processing training programme provided internal competency based training for the recruitment of individuals with less experience. Lastly, an understudy development programme was in place, as all non-Namibian employees had an appointed understudy. Through the Mathematics Support and Enrichment Programme, the mine helped gifted learners to reach their full academic potential. It also supported the Annual National Mathematics Congress, which targets the development of mathematics, as well as the mathematics teaching skills of teachers, across Namibia. Financially disadvantaged, but academically able, learners were also supported through the Mondesa Youth Opportunities Trust.

I.2.3.9. *Site and supporting facilities (infrastructure)*

The mine is located about 90 km east of Walvis Bay, at the foot of the Langer Heinrich Mountain in the Namib Desert and within the Namib-Naukluft National Park. The mine site encompasses two mining licences (ML 140 and ML 172) and accessory works areas of about 74 km^2, of which 4 km^2 has been used to date for mining, waste disposal and processing activities.

The site is connected to the Namibian road infrastructure through a 25 km long access road joining the C28 regional road, which passes the mine in the south-east. The C28 is an unsurfaced gravel road connecting Windhoek with Swakopmund via the Bosua Pass. The site has existing water pipeline and power connections provided by the Namibian water and power utilities, respectively. The water pipeline and related infrastructure run alongside the C28 for about 50 km and then branch off to follow the access road to the site. The section of the water pipeline adjacent to the C28 is located above ground, while the section adjacent to the access road is underground. The power line infrastructure to the mine runs from the Kuiseb substation in Walvis Bay to the access road. From there, it runs parallel to the access road to the mine site. There is also an above ground water pipeline and associated gravel track between the Swakop River boreholes and the mining lease running alongside the Langer Heinrich Mountain towards the operations area.

The mine is surrounded by the Namib-Naukluft National Park; the nearest boundary of the Park to the north of the mine is about 15 km away. This boundary also indicates the location of the nearest commercial farm (i.e. the nearest neighbour), Modderfontein. The northern parts of the park include large tracts of land without any access by road. A small piece of land within the park, close to one of the Swakop River abstraction boreholes, is privately owned. This land is referred to as Farm Riet and, although it can never be developed, the owner has access to the land for camping and other non-intrusive activities.

I.2.3.10. *Contingency planning*

The health environment radiation management system, contingency plans and programmes are in place for all areas and processes requiring such response or contingency planning.

I.2.3.11. *Waste management and minimization*

Waste generated during the different phases of the operations was categorized into mineralized waste and non-mineralized waste and dealt with in terms of a formal waste management and mitigation plan.

Mineralized waste includes all mineralized material that cannot be processed further, because of constraints related to current metallurgical technology and processes, the current low uranium commodity price or both. Mineralized waste is further categorized into mined mineralized waste rock and processed mineralized waste.

Non-mineralized waste includes: low level radioactive contaminated waste stored on-site at the designated waste location; general waste disposed of at the Swakopmund landfill site; hazardous waste taken to the Walvis Bay hazardous waste facility; recyclable material and recyclable metal sold to scrap metal dealers; and medical waste incinerated at the Swakopmund Cottage Medi-Clinic. Only non-mineralized waste with a radioactive surface contamination of less than 4 Bq cm^{-2} was cleared and authorized for removal from the site.

I.2.3.12. Industrial involvement, including procurement

The Langer Heinrich uranium mine is a member of the Namibian Chamber of Mines, and a founding member of the NUA. This enables the mine to contribute to and participate in working groups to address key industry issues and understand the broader challenges of the uranium sector. According to general practice in the Namibian mining industry, procurement of goods and services is done locally within Namibia as far as possible.

I.2.4. Trekkopje mine

The Trekkopje mine is 100% owned by Orano Mining Namibia, a subsidiary of the French Orano Group (previously AREVA). Orano Mining Namibia also owns the Erongo desalination plant, built to supply the Trekkopje mine with water for its operations. Trekkopje deposits (Klein Trekkopje and Trekkopje) are surficial deposits (80% of the mineralization is contained in the top 15 m) hosted in calcrete conglomerate of the Cenozoic era. On 1 January 2017, recoverable resources amounted to 18 720 t U at an average grade of 0.012% U [2]. The mine is situated 70 km north-east of Swakopmund and to the north of the Rössing mine in the Erongo region. The mine site encompasses a licensed mining facility and accessory works areas of about 374 km^2, of which only a small area is used for mining, waste disposal and processing.

Since 2005, the calcrete hosted uranium deposit has been developed in several phases. Mining was performed by blasting, loading and hauling from the open pit, before the uranium-bearing rock was processed and subjected to an alkaline heap leach to produce sodium diuranate. In mid-2013, the almost completed processing plant and associated mining facilities were put in care and maintenance owing to the unfavourable spot price of uranium. A care and maintenance team currently protects the mine's infrastructure so that it can be recommissioned when economic conditions become more favourable.

I.2.4.1. Stakeholder involvement

Orano Mining Namibia has engaged with all stakeholders, including the Namibian Government, at the local, regional and national levels in the areas of economic development, education, culture and sport in the Erongo region. The company continues to support initiatives in these fields, although the mine is not generating income at present.

As the owners of the Erongo desalination plant, Orano Mining Namibia has a slightly different set of stakeholders from the other uranium mines in Namibia. The desalination plant currently supplies the Namibian water utility, NamWater, which in turn provides water to other mining operations and the coastal towns.

Orano Mining Namibia also has a stakeholder partnership with the Mineral Processing Department of the Namibian University of Science and Technology, where joint metallurgical research has been conducted. Orano Mining Namibia's corporate communications officer interacts with all stakeholders to keep them informed about developments at both the mine and the desalination plant.

I.2.4.2. Safety and radiation protection

Orano has a shared safety culture that helps to reduce risks and prevent accidents. Under care and maintenance, Orano Mining Namibia uses its established safety management system to remind employees and contractors at every level of their responsibility for safety, to strengthen risk assessment and prevention, to standardize procedures and share best practices, and to carry out safety campaigns and monitor performance. On 3 October 2018, Orano Mining Namibia achieved a new safety milestone of six continuous years without lost time injury.

Occupational health monitoring includes regular medical examinations at the Chief Medical Officer's practice and continuous radiation monitoring at the mine, which consists of personal and area monitoring. Radiation monitoring results have shown that there is no exposure higher than the background radiation on the mine site. Orano Mining Namibia has an approved radiation management plan that specifies the monitoring requirements and submits annual reports on radiation management to the NRPA.

I.2.4.3 Environmental protection

The environmental performance of Orano Mining Namibia is monitored through internal tracking of environmental indicators, such as water, electricity and fuel consumption, greenhouse gas emissions and production of waste. To check compliance with the environmental management plan and legal requirements, an independent audit of the mine and the Erongo desalination plant is carried out every year. Biannual reports on the status of the environment and on water management are submitted to the Ministry of Environment and Tourism and the Ministry of Agriculture, Water and Forestry. In addition to these reports, the ministries carry out ad hoc inspections to assess the environmental situation at the site. Monitoring of fauna and flora and of restoration trial areas continues during the care and maintenance phase.

I.2.4.4. Protection/enhancement of cultural, tourism, farming, pastoral and similar interests

While there are no farming interests in the desert environment of the Trekkopje mine, Orano Mining Namibia supports social projects in the areas of economic development, education, culture and sport in the neighbouring communities of Arandis, Swakopmund, Spitzkoppe and Usakos and in the wider Erongo region. Microloans to small and medium scale enterprises, bursaries and promotion of safety at schools and sports events are some of the tools used recently. Orano Mining Namibia also contributes to the protection of the Wlotzkasbaken lichen field in the Dorob National Park.

I.2.4.5. Funding and financing

The Trekkopje project is fully financed by the Orano Group of France.

I.2.4.6. Safeguards and security

Namibia is a signatory to all the relevant international instruments for safeguards and security of nuclear material, which also cover the operations of Orano Mining Namibia. Regular reporting to the Ministry of Mines and Energy is carried out and NRPA inspections take place regularly.

I.2.4.7. Transport/export route

As there is no production during the care and maintenance phase, no transport of any product takes place at present. In the event of future production, the final product will be transported by road to the harbour of Walvis Bay for onward shipping in accordance with security and safeguards requirements.

I.2.4.8. Human resources development

Orano Mining Namibia believes that promoting education and skills is indispensable for the development of the country. The availability of well educated and skilled people ensures the long term sustainability of the mining industry and other businesses. The care and maintenance phase has provided employees with more time to upgrade their skills or qualifications by attending training courses on and off the mine site. Identified talents are enrolled in professional or leadership development programmes that are preparing them for roles of increased responsibility.

I.2.4.9. Site and supporting facilities (infrastructure)

The mine is located 70 km north-east of Swakopmund and to the north of the Rössing mine in the Erongo region. The mine site encompasses a mining licence and accessory works area of about 374 km^2, of which only a small area has been used for mining, waste disposal and processing. At present, the existing infrastructure comprises a full recovery plant as well as equipment intended to serve the heap leach pads. All infrastructure is protected and kept in working condition during the care and maintenance phase.

The site is connected to the Namibian road infrastructure through a 23 km long access road, which links up with the B2 Trans-Kalahari Highway about 70 km east of Swakopmund at Arandis. For potential future developments, a rail connection will also be available. The mine is connected to the existing power grid. An independent power producer has established a 5 MW photovoltaic plant on the Trekkopje site.

Orano Mining Namibia owns the largest reverse osmosis seawater desalination plant in southern Africa. It is located near Wlotzkasbaken, 30 km north of Swakopmund, and was inaugurated in 2010. The desalination plant can produce up to 20 million m^3 of salt-free water per year, although its capacity is not fully used while the mine is under care and maintenance. Instead, the plant currently supplies up to 12 million m^3 per annum to the national water utility, NamWater, which is utilized by other mining operations and the coastal municipalities. The mine is linked to the desalination plant through a 40 km long purpose built pipeline. Orano Mining Namibia also has a town office in Swakopmund.

I.2.4.10. Contingency planning

Under the structure of Orano Mining Namibia's HSE management system, contingency plans are in place for all areas and processes requiring emergency planning.

I.2.4.11. Waste management and minimization

Orano Mining Namibia produces limited amounts of waste during the current care and maintenance phase. Under the environmental management plan, this waste is classified into three categories: hazardous waste, non-hazardous waste and recyclable waste, and each category is managed accordingly. Once in operation, all spent ore will be returned to the pit, and approximately 30% of the waste and overburden can also be deposited in the pit. The heap leach method coupled with backfilling has provided a unique opportunity to design the entire operation to reduce the surface footprint of the mine and improve the prospects for post-mining rehabilitation.

I.2.4.12. Industrial involvement, including procurement

Orano Mining Namibia is a Class B Member of the Namibian Chamber of Mines and a founding member of the NUA. The company makes substantial contributions to these two bodies.

According to general practice in the Namibian mining industry, procurement is done locally as far as possible. In 2018, Orano Mining Namibia made 99% of all purchases locally, spending NAD 206 million (including payments for utilities).

Appendix II

CASE STUDY OF THE UNITED REPUBLIC OF TANZANIA

II.1. BACKGROUND OF THE URANIUM INDUSTRY IN THE UNITED REPUBLIC OF TANZANIA

II.1.1. Overview of uranium geology and exploration activity

The geological setting of the United Republic of Tanzania (Tanzania) is favourable for the occurrence of most major metals, hydrocarbons, coal, uranium, phosphate, and metallic and many non-metallic minerals. The geological environment of the country covers the entire chrono-stratigraphic units from the Archaean eon to the Quaternary period [100, 101].

The occurrence of uranium in Tanzania was reported for the first time in 1953 in pegmatites of the Uluguru Mountains in the Morogoro region in the eastern part of the country from locally extracted uraninite. Between 1978 and 1983, the Government of the United Republic of Tanzania sponsored airborne radiometric surveys of the entire country.

Uranium occurs in seven geological types: four sedimentary rock types and three alkaline volcanic rock types. The first uranium mineralization is connected to the Upper Proterozoic stratiform copper mineralization of the Zambian Copperbelt type at Chimala in the southern part of Tanzania, close to the Zambian border. The sandstone type lies on the fluvial Karoo Supergroup, which extends from the great Karoo system from South Africa. High uranium concentration occurs in the Upper Tertiary to Quaternary periods in the surficial (calcrete) type deposit in Bahi and Manyoni in Central Tanzania and the lacustrine phosphate deposit of Minjingu. Uranium is also hosted in rift related alkaline volcanic formations with uraniferous carbonatites, such as Galapo (northern Tanzania) and Panda Hill (southern Tanzania). The uranium occurrence blocks are shown in Fig. 7.

Airborne geophysical surveying and ground follow-up of numerous radiometric anomalies were completed by the Germany based company Uranerzbergbau GmbH [101] from 1978 to 1982. A regional airborne survey and ground follow-up work indicated that blocks A and B in Fig. 7 were the most promising areas for uranium exploration. Uranium mineralization associated with Karoo sandstones was discovered in block A and calcrete related secondary mineralization associated with Mbuga was detected in block B. The unconformity between the Karagwe–Ankolean and Bukoban systems of blocks C and Q appeared to be less prospective for a potential vein-like type uranium deposit

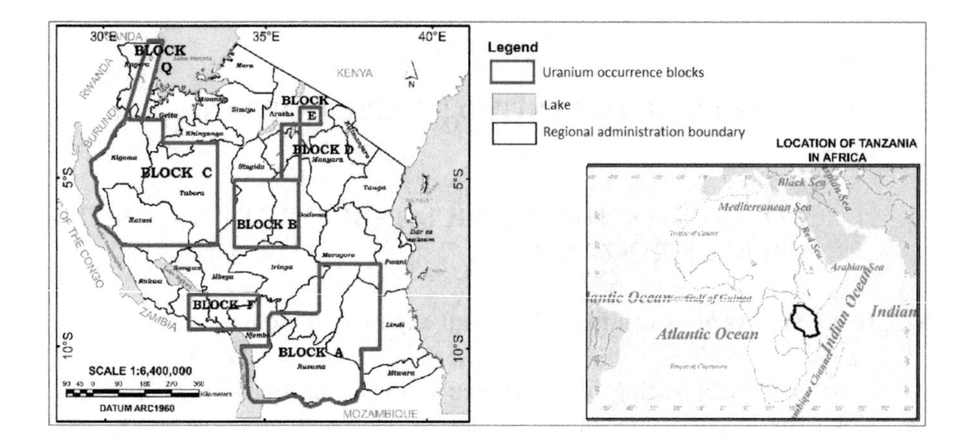

FIG. 7. Uranium occurrences in the United Republic of Tanzania (adapted with permission from the Geological Survey of Tanzania). Block A: Karoo sandstone, vein-like (unconformity); block B: Dodoma calcrete; block C: sandstone; block D: Minjingu and Galappo sedimentary (phosphate), intra-intrusive (carbonatite); block E: Monduli–Tarosero volcanics; block F: Mbeya–Njombe sedimentary (black shales), intra-intrusive (carbonatite), vein-like (unconformity); block Q: Bukoba vein-like (unconformity), intra-intrusive (granite), sandstone.

than the Ubendian/Bukoban unconformity. Block Q and part of block C were classified as uneconomical, while the remaining part of block C was assumed to be potentially economical. Uranium in phosphate was confirmed in block D, but the erratic grades and low tonnage did not justify further exploration. Exploration for uranium in acid volcanics and carbonatites (blocks D, E and F) and uranium in Upper Proterozoic shales (block F) was discontinued owing to low uranium levels in the trachytic basalts of Monduli Juu, the carbonatites of Galappo and Panda and the copper-bearing shales of Chimala.

Blocks A and B showed significant uranium mineralization, in keeping with the sandstone type in the Karoo system, which extends from southern Africa to Tanzania. In addition, these blocks contain calcrete type uranium mineralization. Block A, a Karoo basin located in the south-eastern part of the country, contained in excess of 3000 m of sediments, which were preserved in several north-north-east–north-east striking half-grabens or other structural basin conditions [102]. These are all intracratonic basins, most of which are filled with terrestrial sediments. Sedimentation commenced with glacigene deposits. These are of Late Carboniferous to Early Permian age and may be equated with other glacial successions in Africa and elsewhere in Gondwana. The glacigene beds are overlain by fluvial–deltaic coal-bearing deposits succeeded by arkoses and continental red beds. In 1982, uranium exploration was stopped because of low uranium prices in the world market.

Interest in uranium exploration was rekindled after the rise of uranium prices in 2007, when the government granted over 70 uranium prospecting licences. The second exploration wave was centred in Manyoni and Bahi in central Tanzania and Mkuju River in southern Tanzania. Uranex NL and TanzOz Uranium Ltd performed extensive exploration in the Bahi and Manyoni uranium exploration sites, while Mantra Resources Ltd concentrated on the Mkuju River basin in the Karoo system in southern Tanzania.

The exploration and feasibility studies in the Bahi, Manyoni and Mkuju River uranium exploration projects identified several areas with potential low grade uranium deposits. Pre-feasibility studies of the Bahi and Manyoni deposits found that low surficial uranium processing was not economical. This finding pushed Uranex to move to the Mkuju River sandstone hosted uranium deposit for further exploration. Mantra Resources Ltd concentrated on the Mkuju River project (MRP), which is located in the Selous Game Reserve — classified as a UNESCO (United Nations Educational, Scientific and Cultural Organization) World Heritage site — in the south of the country.

II.2. MILESTONES ASSOCIATED WITH THE URANIUM PROJECT IN THE UNITED REPUBLIC OF TANZANIA

The uranium project at Mkuju River conforms to Milestone 2 in the uranium production cycle (i.e. a country proposing to initiate or reinvigorate uranium mining, with known exploitable reserves). Mantra/Uranium One (see Section II.2.5) completed a definitive feasibility study at the end of 2013 and was granted a special mining licence (SML) to commence uranium mining. Historically, an SML is granted for large scale mining operations with a capital investment of more than $100 million. The designed production approved by the SML is 2300 t U per year.

II.2.1. Mkuju River project viability

The project was declared economically viable in early 2011 after the completion of a definitive feasibility study (DFS).

II.2.2. Stakeholder involvement

Mantra/Uranium One has won strong support from the Government of the United Republic of Tanzania because of the commitment of the operator to cooperate on social responsibility requirements during uranium exploration. The operator also benefits from local support from the nearby village of

Likuyu Sekamaganga and other villages surrounding the proposed mine area. The support is attributable to a well developed corporate social responsibility programme that is sensitive to local needs and addresses various stakeholder groups with targeted activities and support. The operator has supported schools, health care, poaching prevention, and small and medium scale enterprise development. During exploration, the operator used local suppliers for procurement — for example, for poultry and other food produce. The operator employed over 90% of the semi-skilled labour from the region. A list of stakeholders of the uranium industry in Tanzania was carefully considered and a detailed stakeholder map was created (Fig. 8) that highlights both direct and indirect stakeholders associated with the industry.

Mantra/Uranium One has established a plan for infrastructure development in Likuyu Sekamaganga and surrounding areas. The planned upgrades to address the infrastructure needs in a collaboration between the operator, the community and the district authorities are as follows:

— Development of a ten year town master plan for the Namtumbo district, which sets out new areas for residential expansion, transport hubs, industrial developments and community facilities, as well as services such as the new district hospital and the expansion of the police station.

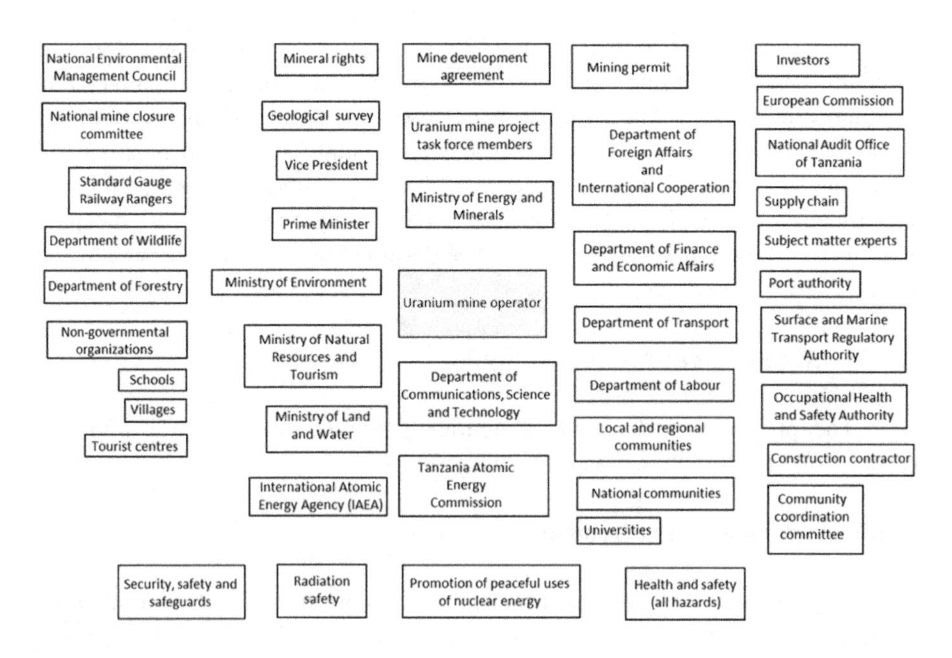

FIG. 8. The United Republic of Tanzania Uranium Project Stakeholder Map, with Project Task Force Membership.

— Government enhancement of the main road from Songea, Namtumbo to Mtwara; plans are in place to upgrade the road between Namtumbo, Likuyu Sekamaganga and the site (with support from Mantra Resources Ltd).
— Installation of water supply improvement projects to service a projected population of 50 000 (from the current population of approximately 20 000).
— Intensification of HIV awareness campaigns, macro- and microloan institutions.
— Development of agricultural projects to supply the mine with fresh food.
— Establishment of a new vocational education centre in Namtumbo.
— Allocation of 150 ha in Likuyu Sekamaganga for the mine village and associated infrastructure.
— Allocation of land for a new police station.
— Development of supporting infrastructures, such as the following:
  (i) Water supply infrastructure;
  (ii) A new hospital and laboratory;
  (iii) Possible future electricity supply.
— Implementation of other initiatives relating to community development, including the following:
  (i) Game scout patrols in the wildlife management area;
  (ii) 500 desks for secondary schools and 360 desks for primary schools in the area;
  (iii) Bursary support for 287 pupils to attend secondary school;
  (iv) Support for microprojects such as sewing of overalls, rearing of chickens, egg production and vegetable gardening.

II.2.3. Safety and radiation protection

The MRP has established a system of safety and radiation protection for workers and visitors in accordance with established regulatory requirements, as stipulated in the Tanzanian national regulatory framework. The operator has established a radiation safety and protection programme in accordance with international guidelines, standards and recommendations to ensure thorough control practices and measures. Personnel, workers and members of the public visiting the site will not be exposed to effective doses exceeding limits recommended by national laws and the International Commission on Radiological Protection as a result of the uranium mining and processing of uranium, surface storage of mined product and processed material at the site at Mkuju [103].

II.2.4. Environmental and social impact assessment

According to the Environmental Management Act 2015, an ESIA is mandatory for large mining projects. This is an important document for building confidence regarding the project's environmental and social impact and is required by the Ministry of Minerals to issue a mining licence. Normally, the owner/operator appoints a company registered by the National Environmental Management Council (NEMC) to perform the ESIA in areas covered by the operator's prospecting licence. The process of conducting an assessment of the possible environmental and social impacts of the project consists of several stages, such as screening, scoping and preparation of terms of reference, conducting the EIA, preparation of the EIS, preparation of the environmental management plan, stakeholder consultations and review. These stages lead to the ESIA report, which is audited by the appointed committee according to the Environmental Management (Impact Assessment and Audit) Regulations 2005. The audit committee report is submitted to the NEMC and is normally used by the NEMC to obtain environmental clearance by the Minister of Environment.

For the MRP, the ESIA [102] was performed in full to ensure that the mining, production and associated activities are carried out in a manner that complies with the locally and internationally accepted best practices in environmental, health, safety and socioeconomic parameters. The objective of the ESIA for the proposed uranium mining was to identify the potential impacts and propose mitigation measures to ensure that the project is implemented according to international environmentally acceptable standards.

II.2.5. Project timeline

The MRP is the most advanced uranium project in the United Republic of Tanzania. It was declared economically viable in early 2011 after a DFS [102]. The history of resource progression is outlined in Table 6 [102].

TABLE 6. MKUJU RIVER URANIUM EXPLORATION AND RESOURCE PROGRESSION

Year	Event
2007	February: Mantra Resources Ltd commences work on-site June: first drill hole on-site completed
2008	Approximately 40 000 m of drilling completed

TABLE 6. MKUJU RIVER URANIUM EXPLORATION AND RESOURCE PROGRESSION (cont.)

Year	Event
2009	January: Maiden resource of 13 800 t U released PFS commenced ~105 000 m drilling completed
2010	Release of updated resource estimate of 32 400 t U DFS initiated by Mantra Resources Ltd ~250 000 m of drilling completed Resource estimate updated to 39 000 t U ARMZ/Uranium One offer to acquire Mantra Resources Ltd
2011	DFS by Mantra Resources Ltd completed ARMZ/Uranium One offer completed Release of updated resource combined to 45 800 t U (measured and indicated and inferred)
2012	Resources updated to 44 600 t U Increased process plant size to 2300 t U per year of production (Mantra DFS) 'Value engineering' exercise commenced UNESCO approval to set site outside the formal boundaries of the Selous Basin and Selous Game Reserve ESIA approval Environmental clearance from Vice President's office; environmental permits issued
2013	April: SML approval by the United Republic of Tanzania for 2 300 t U per year Consent to operate in the game reserve in progress Uranium One DFS initiated (completed in 2013) Resource estimate updated to 58 460 t U
2016	Mining Development Agreement negotiation suspended Uranium market depressed Application for maintenance and care

Note: PFS: pre-feasibility study; DFS: definitive feasibility study; ESIA: environmental and social impact assessment; SML: special mining licence.

In 2012, Uranium One acquired a minority share in the MRP and became the operator of Mantra Resources Tanzania Ltd. Mantra/Uranium One acquired an SML from the Ministry of Minerals in 2013. Construction was planned to commence in early 2014 and operation in 2015. However, a drop in the spot price of uranium in the market forced the company to postpone the plan because the project became uneconomical.

Since 2013, the company has been maintaining the mine by continuing a reduced level of exploration activity (Fig. 9) and retrenching staff. In 2016, the company applied for a permit for care and maintenance from the Ministry of Minerals. The mine has followed the same trend as a number of uranium development projects across Africa that were frozen and postponed, including the Trekkopje (Tanzania), Langer Heinrich (Namibia), Imouraren (Niger), Somnia (Niger) and Kayelekera (Malawi) mines.

II.2.6. Anticipated economic impact of uranium project

As most global uranium producers, Tanzania does not have production facilities to support the complete nuclear fuel cycle (e.g. mining to conversion) and does not intend to develop this processing capability. The domestically produced UOC will be solely for export to its current owner or to the world market, in conformity with national and international laws and safeguards. Uranium is expected to generate benefits through payment of government taxes,

FIG. 9. Mkuju River exploration site (courtesy of D.A. Mwalongo, Tanzania Atomic Energy Commission).

royalties, increase of the employment rate and stimulation of the local economy, in particular within Namtumbo district. Skills development will be a benefit both in the exploitation of uranium and through fully transferable skills, such as natural resource project planning, management and regulation, stakeholder engagement, ESIAs and strategic planning for sustainable development. In due course, the MRP will be the first uranium mine in Tanzania and will rank among the top ten uranium deposits in the world. It is estimated that it will employ more than 700 workers when operation starts. Mantra/Uranium One has already invested over $211 million on project development. An additional $700 million is required for further development.

II.2.7. Government policy for uranium mining in the United Republic of Tanzania

The mining industry in Tanzania is governed by the minerals policy of 2009. The policy is anchored to the role of minerals in achieving the goal of a sustainable, integrated economy in the coming 25 years. The aim of this policy is to continue to attract private companies to take the lead in exploration, mining, mineral beneficiation and marketing with the purpose of increasing the mineral sector's contribution to the national GDP and reducing poverty by integrating the mining industry into the national economy. Other objectives declared in the policy are as follows:

(a) Improve the economic environment in order to attract and sustain local and international private investment in the mineral sector;
(b) Promote economic integration between the mineral sector and other sectors of the economy;
(c) Strengthen the legal and regulatory framework of the mineral sector and enhance the capacity for monitoring and enforcement;
(d) Strengthen the institutional capacity for effective administration and monitoring of the mineral sector;
(e) Participate strategically in viable mining projects and establish an enabling environment for Tanzanians to participate in ownership of medium and large scale mines;
(f) Support and promote the development of small scale mining so as to increase its contribution to the economy;
(g) Establish transparent and adequate land compensation, relocation and resettlement schemes in mining operations;
(h) Strengthen the involvement and participation of local communities in mining projects and encourage mining companies to increase their corporate social responsibility commitments;

(i) Promote and facilitate value addition activities within the country to increase income and employment opportunities;
(j) Promote the research, development and training required in the mineral sector and encourage their utilization;
(k) Improve communication concerning the mineral sector with respect to the public through education and provision of accurate and timely information.

In early July 2017, the Government of the United Republic of Tanzania revised the mining laws to foster mine development, allowing the country to benefit from its rich mineral wealth. Mineral royalty rates were separated into two categories. For gemstones and diamonds, the royalty rate increased from 5% to 6%. For metallic minerals such as gold, silver and copper, the royalty rate also increased from 4% to 6%. In the Mining Act No. 14 of 2010 [104], uranium falls under energy minerals such as coal, for which the royalty rate is similar to metallic minerals, at 6%.

The new regulations empower the government to renegotiate existing agreements and make more advantageous new ones. The income legislation supersedes all other laws, such as the Mining Act, which provides for fiscal stability clauses. The revised laws are flexible to allow the Government to convert tax expenditures resulting from stability agreements into equity holdings in a mining operation. The laws possess robust measures for the training of local staff. It also defines local stakeholders, requirements for insurance coverage, strict liability for environmental damage and provision for cooperative social responsibility expenditure.

II.2.8. National position on the milestones

The United Republic of Tanzania's mining project at Mkuju River falls under Milestone 2 in the uranium production cycle (i.e. the country proposes to initiate or reinvigorate uranium mining, with known exploitable reserves). Mantra/Uranium One completed a definitive feasibility study at the end of 2013 and were granted an SML to commence uranium mining. Historically, an SML is granted for large scale mining operations with capital investment of more than $100 million. The intended production as approved by the SML is 2300 t U per year.

II.2.9. Uranium mining regulatory framework

II.2.9.1. Mining legislation and regulatory framework

According to the Mining Act No. 14 of 2010 [104] and the Atomic Energy Act No. 7 of 2003 [105], the Ministry of Minerals was established as the main regulator for all mining and related activities. The Mining Act set up the legal framework governing mineral exploration and exploitation in Tanzania, advocating the general principles, authorization requirements, administrative measures, royalties, fees and other charges applicable to mining activities, the reporting requirements and other related provisions. However, it also regulates radioactive minerals, including uranium, which are classified as 'energy minerals'. These are defined in section 4(1) as "a group of minerals comprising of coal, peat, uranium, thorium and other radioactive minerals". Radioactive minerals are defined in Section 108(4) [104].

The Ministry of Minerals regulated uranium mining by issuing a prospecting licence and later an SML that allows the operator to start uranium mining and processing. Further, on behalf of the Government, it negotiates and agrees with the operator on how the Government will be compensated on the basis of royalties and taxes. This is defined in the Mining Development Agreement.

II.2.9.2. Environmental legislation and regulatory framework

Environmental regulatory control for the uranium mining industry is governed by the Environmental Management Act, 2004. The law establishes the legal and institutional framework for sustainable management of the environment, stating the principles for environmental management and the requirements for impact assessment, prevention and control of pollution, waste management, environmental quality standards, public participation, compliance and enforcement. The objective of the Environmental Management Act is to provide for and promote the enhancement, protection, conservation and management of the environment.

The NEMC issues environmental clearance if the project qualifies after an ESIA has been performed and audited. The NEMC also performs routine environmental inspections to ensure that the mines conform to environment requirements and standards.

II.2.9.3. Nuclear legislation and regulatory framework

The Atomic Energy Act established the Tanzania Atomic Energy Commission as the nuclear regulatory body, prescribing its functions in relation to the control of ionizing radiation to protect the people and the environment from the harmful effects of ionizing radiation. It was revised in 2017 to incorporate the updated international safety standards and the IAEA recommendations made by the 2013 Uranium Production Site Appraisal Team mission and the 2015 Integrated Regulatory Review Service mission [79]. The Act has been revised in line with the national policies on nuclear law. The Atomic Energy (Radiation Safety in the Mining and Processing of Radioactive Ores) Regulations of 2011 provide the regulatory framework for radiation safety in the uranium mining industry. The regulations include provisions for radioactive waste management, including guidelines for the development of a radioactive waste management plan by the licence holders, the observance of the Radioactive Waste Management for the Protection of Human Health and Environment Regulations, and the storage of radioactive waste and tailings from processing facilities designed and constructed to offer maximum containment. Other regulations that are under revision for the control of the uranium mining life cycle include the following:

(a) The Atomic Energy (Radioactive Waste Management) Regulations of 2019;
(b) The Atomic Energy (Protection from Ionizing Radiation) Regulations of 2019;
(c) The Atomic Energy (Packaging and Transport of Radioactive Material) Regulations of 2019;
(d) The Atomic Energy (Security of Radioactive Sources) Regulations of 2019;
(e) The Atomic Energy (Security of Nuclear Material and Facilities) Regulations of 2019.

REFERENCES

[1] INTERNATIONAL ATOMIC ENERGY AGENCY, Milestones in the Development of a National Infrastructure for Nuclear Power, IAEA Nuclear Energy Series No. NG-G-3.1 (Rev. 1), IAEA, Vienna, Austria (2015).

[2] OECD NUCLEAR ENERGY AGENCY, INTERNATIONAL ATOMIC ENERGY AGENCY, Uranium 2018: Resources, Production and Demand, OECD, Paris (2019) 435–438.

[3] INTERNATIONAL ATOMIC ENERGY AGENCY, Quantitative and Spatial Evaluations of Undiscovered Uranium Resources, IAEA-TECDOC-1861, IAEA, Vienna (2018).

[4] INTERNATIONAL ATOMIC ENERGY AGENCY, Geological Classification of Uranium Deposits and Description of Selected Examples, IAEA-TECDOC-1842, IAEA, Vienna (2018).

[5] OECD NUCLEAR ENERGY AGENCY, Forty Years of Uranium Resources, Production and Demand in Perspective: The Red Book Retrospective, OECD, Paris (2006).

[6] INTERNATIONAL ATOMIC ENERGY AGENCY, World Distribution of Uranium Deposits (UDEPO), IAEA-TECDOC-1843, IAEA, Vienna (2018).

[7] INTERNATIONAL ATOMIC ENERGY AGENCY, World Uranium Geology, Exploration, Resources and Production, IAEA, Vienna (2020).

[8] INTERNATIONAL ATOMIC ENERGY AGENCY, Best Practice in Environmental Management of Uranium Mining, IAEA Nuclear Energy Series No. NF-T-1.2, IAEA, Vienna (2010).

[9] INTERNATIONAL ATOMIC ENERGY AGENCY, Guidebook on Environmental Impact Assessment for In Situ Leach Mining Projects, IAEA-TECDOC-1428, IAEA, Vienna (2005).

[10] INTERNATIONAL ATOMIC ENERGY AGENCY, Environmental Impact Assessment for Uranium Mine, Mill and In Situ Leach Projects, IAEA-TECDOC-979, IAEA, Vienna (1997).

[11] COMMITTEE FOR MINERAL RESERVES INTERNATIONAL REPORTING STANDARDS, International Reporting Template for the Public Reporting of Exploration Results, Mineral Resources and Mineral Reserves, ICMM, London (2006).

[12] ONTARIO SECURITIES COMMISSION, Standards of Disclosure for Mineral Projects, National Instrument NI-43 101, OSC, Toronto (2011).

[13] JOINT ORE RESERVES COMMITTEE, Australasian Code for Reporting of Exploration Results, Mineral Resources and Ore Reserves, JORC, Carlton (2012).

[14] SAMCODES STANDARDS COMMITTEE, The South African Code for the Reporting of Exploration Results, Mineral Resources and Mineral Reserves (the SAMREC Code), Marshalltown (2016).

[15] PELIZZA, M.S., BARTELS, C.S., "Introduction to uranium in situ recovery technology", Uranium for Nuclear Power: Resources, Mining and Transformation to Fuel (HORE-LACY, I., Ed.), Woodhead Publishing, Cambridge (2016) 157–213.

[16] INTERNATIONAL ATOMIC ENERGY AGENCY, In Situ Leach Uranium Mining: An Overview of Operations, IAEA Nuclear Energy Series No. NF-T-1.4, IAEA, Vienna (2016).
[17] INTERNATIONAL ATOMIC ENERGY AGENCY, Manual of Acid in Situ Leach Uranium Mining Technology, IAEA-TECDOC-1239, IAEA, Vienna (2001).
[18] INTERNATIONAL ATOMIC ENERGY AGENCY, Prospective Radiological Environmental Impact Assessment for Facilities and Activities, IAEA Safety Standards Series No. GSG-10, IAEA, Vienna (2018).
[19] BULLOCK, R.L., MERNITZ, S., Mineral Property Evaluation, Handbook for Feasibility Studies and Due Diligence, Society for Mining, Metallurgy and Exploration, Englewood (2018).
[20] WOODS, P.H., "Uranium mining (open cut and underground) and milling", Uranium for Nuclear Power: Resources, in Mining and Transformation to Fuel (HORE-LACY, I., Ed.), Woodhead Publishing, Cambridge (2016) 125–156.
[21] INTERNATIONAL ATOMIC ENERGY AGENCY, Establishment of Uranium Mining and Processing Operations in the Context of Sustainable Development, IAEA Nuclear Energy Series No. NF-T-1.1, IAEA, Vienna (2009).
[22] INTERNATIONAL ATOMIC ENERGY AGENCY, Steps for Preparing Uranium Production Feasibility Studies: A Guidebook, IAEA-TECDOC-885, IAEA, Vienna (1996).
[23] INTERNATIONAL ATOMIC ENERGY AGENCY, Safety Assesment for Facilities and Activities, IAEA Safety Standard Series No. GSR Part 4 (Rev. 1), IAEA, Vienna (2016).
[24] INTERNATIONAL ATOMIC ENERGY AGENCY, Decommissioning of Facilities, IAEA Safety Standards Series No. GSR Part 6, IAEA, Vienna (2014).
[25] COLLIER, D., "Uranium mine and mill remediation and reclamation", Uranium for Nuclear Power: Resources, Mining and Transformation to Fuel (HORE-LACY, I., Ed.), Woodhead Publishing, Cambridge (2016) 415–437.
[26] OECD NUCLEAR ENERGY AGENCY, INTERNATIONAL ATOMIC ENERGY AGENCY, Environmental Remediation of Uranium Production Facilities, OECD, Paris (2002).
[27] INTERNATIONAL ATOMIC ENERGY AGENCY, IAEA Safety Glossary: 2018 Edition, IAEA, Vienna (2019).
[28] EUROPEAN COMMISSION, FOOD AND AGRICULTURE ORGANIZATION OF THE UNITED NATIONS, INTERNATIONAL ATOMIC ENERGY AGENCY, INTERNATIONAL LABOUR ORGANIZATION, OECD NUCLEAR ENERGY AGENCY, PAN AMERICAN HEALTH ORGANIZATION, UNITED NATIONS ENVIRONMENT PROGRAMME, WORLD HEALTH ORGANIZATION, Radiation Protection and Safety of Radiation Sources: International Basic Safety Standards, IAEA Safety Standards Series No. GSR Part 3, IAEA, Vienna (2014).
[29] INTERNATIONAL ATOMIC ENERGY AGENCY, Occupational Radiation Protection, IAEA Safety Standards Series No. GSG-7, IAEA, Vienna (2018).

[30] EUROPEAN ATOMIC ENERGY COMMUNITY, FOOD AND AGRICULTURE ORGANIZATION OF THE UNITED NATIONS, INTERNATIONAL ATOMIC ENERGY AGENCY, INTERNATIONAL LABOUR ORGANIZATION, INTERNATIONAL MARITIME ORGANIZATION, OECD NUCLEAR ENERGY AGENCY, PAN AMERICAN HEALTH ORGANIZATION, UNITED NATIONS ENVIRONMENT PROGRAMME, WORLD HEALTH ORGANIZATION, Fundamental Safety Principles, IAEA Safety Standards Series No. SF-1, IAEA, Vienna (2006).

[31] INTERNATIONAL ATOMIC ENERGY AGENCY, Governmental, Legal and Regulatory Framework for Safety, IAEA Safety Standards Series No. GSR Part 1 (Rev. 1), IAEA, Vienna (2016).

[32] INTERNATIONAL ATOMIC ENERGY AGENCY, Handbook on Nuclear Law: Implementing Legislation, IAEA, Vienna (2010).

[33] INTERNATIONAL NUCLEAR SAFETY ADVISORY GROUP, Independence in Regulatory Decision Making, INSAG-17, IAEA, Vienna (2003).

[34] OECD NUCLEAR ENERGY AGENCY, Perceptions and Realities in Modern Uranium Mining, OECD, Paris (2014).

[35] INTERNATIONAL ATOMIC ENERGY AGENCY, Establishing the Nuclear Security Infrastructure for a Nuclear Power Programme, IAEA Nuclear Security Series No. 19, IAEA, Vienna (2013).

[36] Treaty on the Non-Proliferation of Nuclear Weapons, INFCIRC/140, IAEA, Vienna (1970).

[37] The Structure and Content of Agreements Between the Agency and States Required in Connection with the Treaty on the Non-Proliferation of Nuclear Weapons, INFCIRC/153, IAEA, Vienna (1972).

[38] Model Protocol Additional to the Agreement(s) between State(s) and the International Atomic Energy Agency for the Application of Safeguards, INFCIRC/540, IAEA, Vienna (1997).

[39] INTERNATIONAL ATOMIC ENERGY AGENCY, Nuclear Material Accounting Handbook, IAEA Services Series No. 15, IAEA, Vienna (2008).

[40] INTERNATIONAL ATOMIC ENERGY AGENCY, Guidance for States Implementing Comprehensive Safeguards Agreements and Additional Protocols, IAEA Services Series No. 21, IAEA, Vienna (2016).

[41] INTERNATIONAL ATOMIC ENERGY AGENCY, Safeguards Implementation Guide for States with Small Quantities Protocols, IAEA Services Series No. 22, IAEA, Vienna (2016).

[42] INTERNATIONAL ATOMIC ENERGY AGENCY, Guidebook on the Development of Regulations for Uranium Deposit Development and Production, IAEA-TECDOC-862, IAEA, Vienna (1996).

[43] INTERNATIONAL ATOMIC ENERGY AGENCY, Model Regulations for the Use of Radiation Sources and for the Management of the Associated Radioactive Waste, IAEA-TECDOC-1732, IAEA, Vienna (2013).

[44] INTERNATIONAL ATOMIC ENERGY AGENCY, Functions and Processes of the Regulatory Body for Safety, IAEA Safety Standards Series No. GSG-13, IAEA, Vienna (2018).

[45] INTERNATIONAL ATOMIC ENERGY AGENCY, Organization, Management and Staffing of the Regulatory Body for Safety, IAEA Safety Standards Series No. GSG-12, IAEA, Vienna (2018).

[46] INTERNATIONAL ATOMIC ENERGY AGENCY, Release of Sites from Regulatory Control on Termination of Practices, IAEA Safety Standards Series No. WS-G-5.1, IAEA, Vienna (2006).

[47] ERNST AND YOUNG, Top 10 Business Risks Facing Mining and Metals in 2019–20, Ernst and Young, London (2018), https://assets.ey.com/content/dam/ey-sites/ey-com/en_gl/topics/mining-metals/mining-metals-pdfs/ey-top-10-business-risks-facing-mining-and-metals-in-2019-20_v2.pdf

[48] INTERNATIONAL ATOMIC ENERGY AGENCY, Communication and Consultation with Interested Parties by the Regulatory Body, IAEA Safety Standards Series No. GSG-6, IAEA, Vienna (2017).

[49] INTERNATIONAL ATOMIC ENERGY AGENCY, Stakeholder Engagement in Nuclear Programmes, IAEA Nuclear Energy Series No. NG-G-5.1, IAEA, Vienna (2021).

[50] INTERNATIONAL ATOMIC ENERGY AGENCY, Communication and Stakeholder Involvement in Environmental Remediation Projects, IAEA Nuclear Energy Series No. NW-T-3.5, IAEA, Vienna (2014).

[51] INTERNATIONAL ATOMIC ENERGY AGENCY, Remediation Strategy and Process for Areas Affected by Past Activities or Events, IAEA Safety Standards Series No. GSG-15, IAEA, Vienna (2020).

[52] INTERNATIONAL ATOMIC ENERGY AGENCY, Monitoring and Surveillance of Radioctive Waste Disposal Facilities, IAEA Safety Standards Series No. SSG-31, IAEA Vienna (2014).

[53] INTERNATIONAL ATOMIC ENERGY AGENCY, INTERNATIONAL LABOUR OFFICE, Occupational Radiation Protection in the Mining and Processing of Raw Materials, IAEA Safety Standards Series No. RS-G-1.6, IAEA, Vienna (2004).

[54] HARRIS, F., "Management for health, safety, environment, and community in uranium mining and processing", Uranium for Nuclear Power: Resources, Mining and Transformation to Fuel (HORE-LACY, I., Ed.), Woodhead Publishing, Cambridge (2016).

[55] INTERNATIONAL ATOMIC ENERGY AGENCY, Assessing the Need for Radiation Protection Measures in Work Involving Minerals and Raw Materials, Safety Reports Series No. 49, IAEA, Vienna (2007).

[56] INTERNATIONAL ATOMIC ENERGY AGENCY, Monitoring and Surveillance of Residues from the Mining and Milling of Uranium and Thorium, Safety Reports Series No. 27, IAEA, Vienna (2002).

[57] INTERNATIONAL ATOMIC ENERGY AGENCY, Occupational Radiation Protection in the Uranium Mining and Processing Industry, Safety Reports Series No. 100, IAEA, Vienna (2020).

[58] INTERNATIONAL ATOMIC ENERGY AGENCY, Disposal of Radioactive Waste, IAEA Safety Standards Series No. SSR-5, IAEA, Vienna (2011).
[59] INTERNATIONAL ATOMIC ENERGY AGENCY, Disposal Options for Disused Radioactive Sources, Technical Reports Series No. 436 IAEA, Vienna (2005), IAEA.
[60] INTERNATIONAL ATOMIC ENERGY AGENCY, Impact of New Environmental and Safety Regulations on Uranium Exploration, Mining, Milling and Management of Its Waste, IAEA-TECDOC-1244, IAEA, Vienna (2001).
[61] MINING ASSOCIATION OF CANADA, A Guide to the Management of Tailings Facilities, Version 3.1, MAC, Ottawa (2019).
[62] MINING ASSOCIATION OF CANADA, Developing and Operation, Maintenance, and Surveillance Manual for Tailings and Water Management Facilities, Second Edition, MAC, Ottawa (2019).
[63] INTERNATIONAL ATOMIC ENERGY AGENCY, Closeout of Uranium Mines and Mills: A Review of Current Practices, IAEA-TECDOC-939, IAEA, Vienna (1997).
[64] INTERNATIONAL ATOMIC ENERGY AGENCY, Environmental Contamination from Uranium Production Facilities and their Remediation (Proc. Int. Workshop Lisbon, 2004), IAEA, Vienna (2005).
[65] THE RÖSSING FOUNDATION, About the Rössing Foundation (1999), https://www.rossingfoundation.com/background.html
[66] KUEPPER, J., How to Use Canada's SEDAR, The Balance (2019), https://www.thebalancemoney.com/how-to-use-canada-s-sedar-1979205
[67] PWC AUSTRALIA, Listing a Company on the Stock Exchange, PwC Australia (2019).
[68] VENTURE LAW CORPORATION, Listing Requirements of the TSX Toronto Venture Exchange (TSX-V) — Exploration and Mining Companies, https://venturelawcorp.com/tsx-venture-exchange-mining-exploration/
[69] JOHANNESBURG STOCK EXCHANGE, Listing Requirements, https://www.jse.co.za/sites/default/files/media/documents/2019-04/JSE%20 Listings%20Requirements.pdf
[70] AUSTRALIA SECURITIES EXCHANGE, Additional Reporting on Mining and Oil and Gas Production and Exploration Activities (2019), https://www.asx.com.au/documents/rules/Chapter05.pdf
[71] AUSTRALIA SECURITIES EXCHANGE, ASX Listing Rules, https://www2.asx.com.au/about/regulation/rules-guidance-notes-and-waivers/ asx-listing-rules-guidance-notes-and-waivers
[72] AUSTRALIA SECURITIES EXCHANGE, Listing Requirements, https://www2.asx.com.au/listings/how-to-list/listing-requirements#:~:text= Admission%20criteria,-Admission%20criteria&text=Your%20company%20must%20 have%20at,the%20listing%20application%20is%20made
[73] BULLOCK, R.L., Accuracy of feasibility study evaluations would improve accountability, Min. Eng. **63** (2011) 78–83.
[74] MARLATT, J., "The business of exploration: Discovering the next generation of economic uranium deposits", Quantitative and Spatial Evaluations of Undiscovered Uranium Resources, IAEA-TECDOC-1861, IAEA, Vienna (2018) 45–96.
[75] DE BEERS, Diamond Exploration, The Insight Report (2014) 46–49.

[76] SCHODDE, R., "Long-term trends and outlook for uranium exploration: Are we finding enough uranium?", Quantitative and Spatial Evaluations of Undiscovered Uranium Resources, IAEA-TECDOC-1861, IAEA, Vienna (2018) 19–44.

[77] GROVES, D.I., SANTOSH, M., Province-scale commonalities of some world-class gold deposits: implications for mineral exploration, Geosc. Front. **6** 3 (2015) 389–399.

[78] INTERNATIONAL ATOMIC ENERGY AGENCY, Nuclear Security in the Uranium Extraction Industry, IAEA-TDL-003, IAEA, Vienna (2016).

[79] INTERNATIONAL ATOMIC ENERGY AGENCY, Nuclear Security Recommendations on Radioactive Material and Associated Facilities, IAEA Nuclear Security Series No. 14, IAEA, Vienna (2011).

[80] INTERNATIONAL ATOMIC ENERGY AGENCY, Management of Residues Containing Naturally Occurring Radioactive Material from Uranium Production and Other Activities, IAEA Safety Standards Series No. SSG-60, IAEA, Vienna (2021).

[81] INTERNATIONAL ATOMIC ENERGY AGENCY, Nuclear Security Assessment Methodologies for Regulated Facilities, IAEA-TECDOC-1868, IAEA, Vienna (2019).

[82] INTERNATIONAL ATOMIC ENERGY AGENCY, Computer Security of Instrumentation and Control Systems at Nuclear Facilities, IAEA Nuclear Security Series No. 33-T, IAEA, Vienna (2018).

[83] INTERNATIONAL ATOMIC ENERGY AGENCY, Conducting Computer Security Assessments at Nuclear Facilities, IAEA-TDL-006, IAEA, Vienna (2016).

[84] INTERNATIONAL ATOMIC ENERGY AGENCY, Preventive and Protective Measures Against Insider Threats, IAEA Nuclear Security Series No. 8-G (Rev. 1), IAEA, Vienna (2020).

[85] INTERNATIONAL ATOMIC ENERGY AGENCY, Regulations for the Safe Transport of Radioactive Material, IAEA Safety Standard Series No. SSR-6 (Rev. 1), IAEA, Vienna (2018).

[86] GREEN, C., International transport of uranium concentrates, Packag. Transp. Storage Secur. Radioact. Mater. **18** (2007) 227–229.

[87] INTERNATIONAL ATOMIC ENERGY AGENCY, Guidebook on the Development of Projects for Uranium Mining and Ore Processing, IAEA-TECDOC-595, IAEA, Vienna (1991).

[88] INTERNATIONAL ATOMIC ENERGY AGENCY, Guidebook on Good Practice in the Management of Uranium Mining and Mill Operations and the Preparation for their Closure, IAEA-TECDOC-1059, IAEA, Vienna (1998).

[89] NAMIBIAN URANIUM INSTITUTE Annual Review 2018, Swakopmund (2019), http://www.namibianuranium.org/wp-content/uploads/2020/10/NUI-Annual-Review-2018.pdf

[90] CHAMBER OF MINES OF NAMIBIA, 2018 Annual Report, Chamber of Mines of Namibia, Windhoek (2019).

[91] NAMIBIAN URANIUM ASSOCIATION, Annual Review 2017, Namibian Uranium Association, Swakopmund (2018).

[92] GEOLOGICAL SURVEY OF NAMIBIA, Strategic Environmental Management Plan (SEMP) for the Central Namib Uranium Mining Province, 2017 Annual Report, Ministry of Mines and Energy, Windhoek (2018).
[93] CHAMBER OF MINES OF NAMIBIA, 2017 Annual Report, Chamber of Mines of Namibia, Windhoek (2018).
[94] GEOLOGICAL SURVEY OF NAMIBIA, Strategic Environmental Management Plan (SEMP) for the Central Namib Uranium Mining Province, 2016 Annual Report, Ministry of Mines and Energy, Windhoek (2017).
[95] GEOLOGICAL SURVEY OF NAMIBIA, Strategic Environmental Management Plan (SEMP) for the Central Namib Uranium Mining Province, 2015 Annual Report, Ministry of Mines and Energy, Windhoek (2016).
[96] GEOLOGICAL SURVEY OF NAMIBIA, Strategic Environmental Management Plan (SEMP) for the Central Namib Uranium Mining Province, 2014 Annual Report, Ministry of Mines and Energy, Windhoek (2015).
[97] GEOLOGICAL SURVEY OF NAMIBIA, Strategic Environmental Management Plan (SEMP) for the Central Namib Uranium Mining Province, 2013 Annual Report, Ministry of Mines and Energy, Windhoek (2014).
[98] GEOLOGICAL SURVEY OF NAMIBIA, Strategic Environmental Management Plan (SEMP) for the Central Namib Uranium Mining Province, 2012 Annual Report, Ministry of Mines and Energy, Windhoek (2013).
[99] GEOLOGICAL SURVEY OF NAMIBIA, Strategic Environmental Management Plan (SEMP) for the Central Namib Uranium Mining Province, 2011 Annual Report, Ministry of Mines and Energy, Windhoek (2012).
[100] GEOLOGICAL SURVEY OF TANZANIA, Minerogenic Map of Tanzania and Explanatory Notes for the Minerogenic Map of Tanzania 1:1,5 M, Geological Survey of Tanzania, Dodoma (2015),
https://gmis-tanzania.com
[101] BIANCONI, F., VOGT, J., Uranium Geology of Tanzania, Borntraeger, Berlin (1987).
[102] MANTRA RESOURCES TANZANIA LTD, personal communication, 2013.
[103] INTERNATIONAL COMMISSION ON RADIOLOGICAL PROTECTION, The 2007 Recommendations of the International Commission on Radiological Protection, Publication 103, ICRP, Oxford (2007).
[104] MINING ACT No. 14 OF 2010, The Government Printer, Dar es Salaam (2010).
[105] ATOMIC ENERGY ACT No. 7 OF 2003, The Government Printer, Dar es Salaam (2003).

ABBREVIATIONS

AEB	Atomic Energy Board of Namibia
ALARA	as low as reasonably achievable
CIM	Canadian Institute of Mining
CRIRSCO	Combined Reserves International Reporting Standards Committee
CSA	comprehensive safeguards agreement
DFS	definitive (or detailed) feasibility study
EIA	environmental impact assessment
EIS	environmental impact statement
ESIA	environmental and social impact assessment
ESIS	environmental and social impact study
HSE	health, safety and environmental
INFCIRC	Information Circular (IAEA)
ISO	International Organization for Standardization
LSA	low specific activity
MRP	Mkuju River Project
NEMC	National Environmental Management Council
NOSA	National Occupational Safety Association
NRPA	National Radiation Protection Authority
NUA	Namibian Uranium Association
NUI	Namibian Uranium Institute
OECD/NEA	OECD Nuclear Energy Agency
SEMP	strategic environmental management plan
SOE	state owned enterprise
SSAC	State system of accounting for and control of nuclear material
UOC	uranium ore concentrate

CONTRIBUTORS TO DRAFTING AND REVIEW

Abbes, N.	Groupe Chimique Tunisien, Tunisia
Blaise, J.R.	International Atomic Energy Agency
Brown, G.	Boswell Capital Corporation, Canada
Dunn, G.	Hydromet Pty Ltd, South Africa
Edson, K.	Consultant, United States of America
Good, C.L.	International Atomic Energy Agency
Hama Siddo, A.	Ministry of Mines and Energy, Niger
Hanly, A.	International Atomic Energy Agency
Hilton, J.	Aleff Group, United Kingdom
Itamba, H.	Ministry of Mines and Energy, Namibia
Lazykina, A.	International Atomic Energy Agency
Lopez, L.	National Atomic Energy Commission (CNEA), Argentina
Moldovan, B.	International Atomic Energy Agency
Mwalongo, D.	Tanzania Atomic Energy Commission, United Republic of Tanzania
Roberts, M.	International Atomic Energy Agency
Schneider, G.	Namibian Uranium Institute, Namibia
Scissions, K.	KHS Solutions, Canada
Woods, P.	International Atomic Energy Agency

Consultants Meetings

Vienna, Austria: 12–14 December 2016; 4–7 September 2017

Structure of the IAEA Nuclear Energy Series*

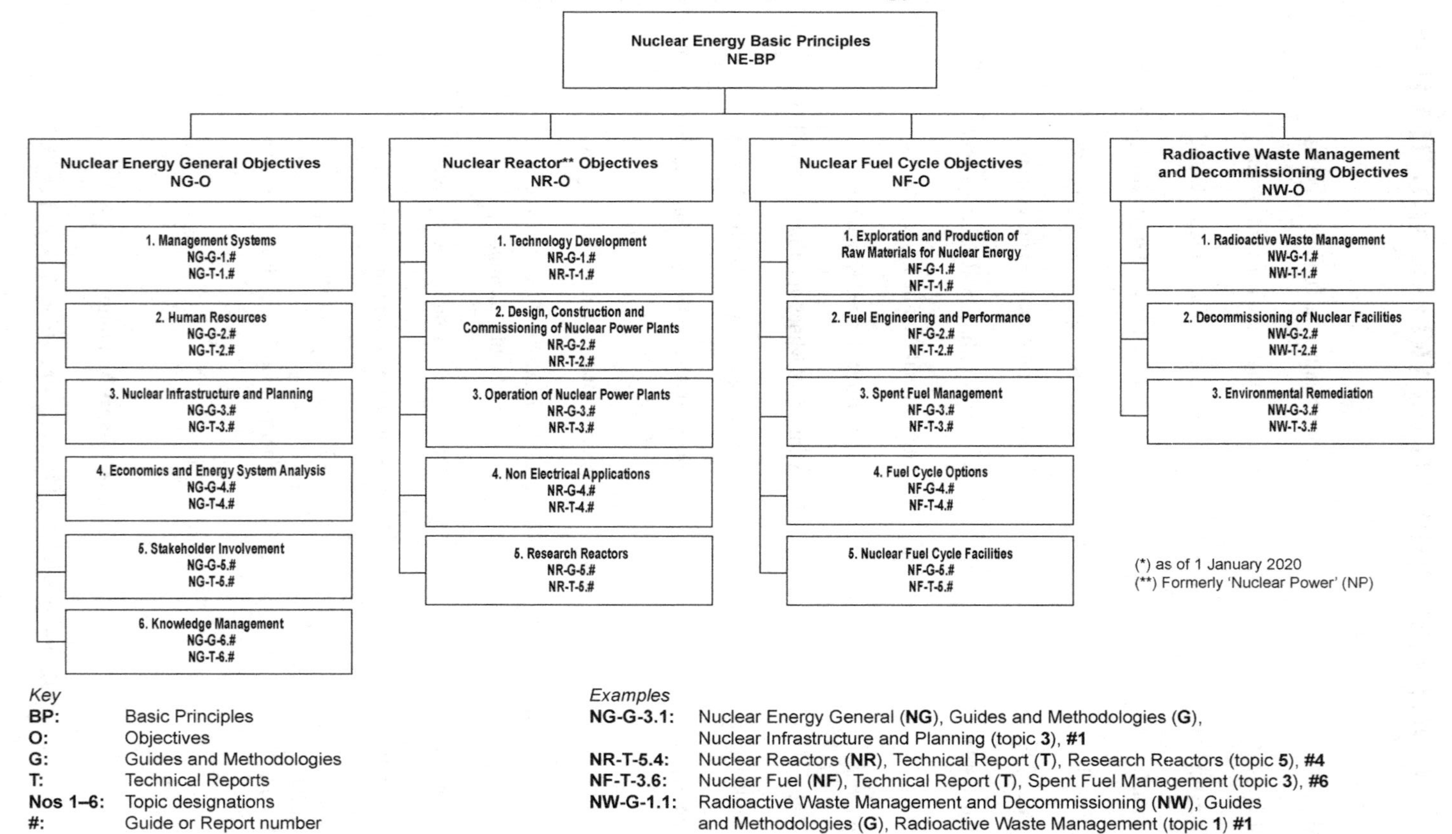

(*) as of 1 January 2020
(**) Formerly 'Nuclear Power' (NP)

Key

BP: Basic Principles
O: Objectives
G: Guides and Methodologies
T: Technical Reports
Nos 1–6: Topic designations
#: Guide or Report number

Examples

NG-G-3.1: Nuclear Energy General (**NG**), Guides and Methodologies (**G**), Nuclear Infrastructure and Planning (topic **3**), **#1**
NR-T-5.4: Nuclear Reactors (**NR**), Technical Report (**T**), Research Reactors (topic **5**), **#4**
NF-T-3.6: Nuclear Fuel (**NF**), Technical Report (**T**), Spent Fuel Management (topic **3**), **#6**
NW-G-1.1: Radioactive Waste Management and Decommissioning (**NW**), Guides and Methodologies (**G**), Radioactive Waste Management (topic **1**) **#1**

No. 26

ORDERING LOCALLY

IAEA priced publications may be purchased from the sources listed below or from major local booksellers.

Orders for unpriced publications should be made directly to the IAEA. The contact details are given at the end of this list.

NORTH AMERICA

Bernan / Rowman & Littlefield
15250 NBN Way, Blue Ridge Summit, PA 17214, USA
Telephone: +1 800 462 6420 • Fax: +1 800 338 4550
Email: orders@rowman.com • Web site: www.rowman.com/bernan

REST OF WORLD

Please contact your preferred local supplier, or our lead distributor:

Eurospan Group
Gray's Inn House
127 Clerkenwell Road
London EC1R 5DB
United Kingdom

Trade orders and enquiries:
Telephone: +44 (0)176 760 4972 • Fax: +44 (0)176 760 1640
Email: eurospan@turpin-distribution.com

Individual orders:
www.eurospanbookstore.com/iaea

For further information:
Telephone: +44 (0)207 240 0856 • Fax: +44 (0)207 379 0609
Email: info@eurospangroup.com • Web site: www.eurospangroup.com

Orders for both priced and unpriced publications may be addressed directly to:

Marketing and Sales Unit
International Atomic Energy Agency
Vienna International Centre, PO Box 100, 1400 Vienna, Austria
Telephone: +43 1 2600 22529 or 22530 • Fax: +43 1 26007 22529
Email: sales.publications@iaea.org • Web site: www.iaea.org/publications